Diatomaceous Earth Filtration for Safe Drinking Water

George P. Fulton

ASCE PRESS
American Society of Civil Engineers
1801 Alexander Bell Drive
Reston, Virginia 20191–4400

Abstract: First adopted by the U.S. Army during World War II, diatomaceous earth (DE) filtration is a viable water treatment method that produces drinking water from low-turbidity sources. This book reviews the range of DE filter equipment, as well as the process performance features of filtration and supplemental treatment units. Performance is evaluated according to the standards set by current and anticipated regulations governing the quality of drinking water. Comparisons are made with other, more traditional methods of water treatment to highlight the particular advantages of DE filtration. Details of comprehensive procedures are presented for the engineer, starting with collection and devleopment of the background data used in preliminary studies and leading to the preparation of final design documents for construction. Site considerations are covered, as are operation and maintenance manuals, including detailed lists of features to consider in troubleshooting DE filter operation faults.

Library of Congress Cataloging-in-Publication Data

Fulton, George P.
Diatomaceous earth filtration for safe drinking water / George P. Fulton.
p. cm.
Includes bibliographical references and index.
ISBN 0-7844-0429-1
1. Water—Purification—Diatomaceous earth filtration. 2. Drinking water—Purification. I. Title.
TD466 .F85 2000
628.1'64—dc21 00-029959

Library of Congress Catalog Card No: 00-029959
ISBN 0-7844-0429-1
Manufactured in the United States of America

Contents

Preface

Diatomaceous earth (DE) filtration has been an effective method for treatment of low turbidity drinking water supplies since World War II. At that time, it was adopted for use for the U.S. Armed Forces. The Army developed the industrial filter for drinking water application because a lightweight, easily transportable filter unit, capable of removing *Entamoeba histolytica* cysts was required. In subsequent tests in the 1980s and after, DE filtration was shown to be the superior alternative for removing *Giardia* and *Cryptosporidium* cysts in the treatment of drinking water.

Despite effective performance, DE filters have had limited use in the treatment of public drinking water supplies in the ensuing period. The several options of rapid sand filter trains using coagulation/flocculation for source water solids conditioning have dominated drinking water treatment since the turn of the century. It is interesting to observe that there has been little change in the basic form of rapid sand filter systems in over a century of use. Certain mechanical features, new materials of construction, and polyelectrolytes to assist coagulation/flocculation have been developed to somewhat change the look of these treatment trains and improve treatment performance. Alum is still the predominant coagulant used to condition solids for removal. In addition to sand, the granular media may also include ground anthracite, a combination of sand and anthracite, and granulated activated carbon.

The academic community has also been reluctant to change sights in the investigation of different methods of drinking water treatment. The elusiveness of the "workings" of the coagulation/flocculation phenomena have provided "meat" for thousands of advanced degree research projects in the past. These investigations still continue and will, no doubt, go on indefinitely in the future in pursuit of the absolute model of the phenomena. On the other hand, diatomaceous earth filtration is a simple, physical separation process for the removal of suspended matter from a flow of drinking water. Robert Bauman produced the definitive work on the operation of DE filters at the Iowa University in the 1960s. There is little left for academic investigation other than DE filtration performance treating different sources of water—certainly not very promising thesis material compared to the pursuit of the elusive rapid sand filter performance model.

As further testament of its efficacy, DE filtration has for many years been the mainstay of the beverage industry in the production of wine and beer. Beverage standards are perhaps more stringent than those for drinking water. This industry use has been responsible for the development of new, improved filtration equipment that may also be used in the treatment of drinking water.

The reader will notice that there are few references at the end of each chapter. This is because very little investigation had been made of DE filtration until the mid-1970s. At that time, the City of New York embarked on a major research program for its Croton watershed supply. DE filtration was a major consideration in these investigations and considerable research data was developed in three successive research programs. The principal performance data used in this text was developed in two of the programs, included in the following reports:

— Metcalf & Eddy, Inc.: Hazen and Sawyer, 1994 Final Report on Development of Design Criteria for the New Croton Water Treatment Plant at Jerome Park Reservoir, Borough of the Bronx, City of New York, February.
— Metcalf & Eddy, Inc.: Hazen and Sawyer, 1999 Final Report on Additional Pilot Studies for the Croton Water Treatment Plant, January.

Little historical background is available about DE filtration from its original use in industry to its present day applications for drinking water and beverage production. This information was pieced together by the author from industrial brochures and interviews with knowledgeable people in the various areas of DE use.

The principal purpose of this text is two-fold:

1. To acquaint those who are unfamiliar with DE filtration or those who have misconceptions about the process with both the advantages and

limitations in using this relatively simple method of treatment in applicable drinking water situations.

2. To provide guidelines for the professional in the development of designs of complete DE facilities and in subsequent start-up and plant operation to produce safe drinking water.

Background theory and practical performance data are provided to assist in the evaluation of DE filtration for different drinking water applications for meaningful comparisons with alternative treatment. Information is provided to permit design of complete DE facilities from material delivery and handling through DE filtration systems to ultimate waste disposal.

Acknowledgments

In today's world of detailed and complex technology, a text of this kind does not result from the endeavors of a single individual. The author is grateful to the City of New York for sponsoring the intensive, comprehensive research program for treatment of the Croton water supply from which a considerable portion of the information used in this book was obtained. In particular, the author recognizes the contributions to that project of members of the Hazen and Sawyer staff: Raymond McIndoe, supervisor of the initial demonstration project: Michael Broder, facilities designer for the demonstration project and subsequent pilot research; and David Rokjer, who dedicated a good portion of a decade to accumulating data on diatomaceous earth filtration performance.

The author also acknowledges with grateful thanks Anita Schleider, for her time and patience devoted to the preparation of the manuscript, as well as Gail Yap and Helen Yoon, whose computer graphics talents are demonstrated by the illustrations included herein.

The author also wishes to thank Michael Elliott of Stearns and Wheler for his peer review of the text. He has served as Chairman of the Precoat Filtration Subcommittee of the AWWA Coagulation and Filtration Committee. In addition to the technical comments, the author is grateful for his advice in moderating the sometimes too ebullient writing style.

The author also thanks both the Celite Corp. and Eagle-Picher Minerals, Inc. for providing background information on the source, mining, manufacturing, delivery, handling, and physical property features of diatomaceous earth media. These data were helpful, not only in the preparation of the text, but also in the design of many of the Chapter 4 graphics.

1
Introduction

With new federal emphasis on pathogenic cyst removal in the Safe Drinking Water Act regulations, the role of the diatomaceous filter in the family of accepted treatment methods takes on more importance. Within limits, the diatomaceous earth filter has superior particulate removal capability, especially in the 2- to 15-microns range of sizes of the predominant *Giardia* and *Cryptosporidum* cysts of concern. Generally, diatomaceous earth (DE) filters are best suited for treatment of source waters with upper turbidity limits from 5 to 10 nephelometric turbidity units (NTU). In most applications, higher turbidity waters would reduce filter runs to a length of time too impracticable for consideration.

Compared to more traditional processes like rapid sand filtration, DE filtration is about the most basic, uncomplicated method available for the removal of discrete particles suspended in a drinking water source as particulates are removed, primarily, simply by entrapment or straining. The liquid flows initially through a precoat of powder-like diatomaceous earth, then through a cake of separated solids accumulating on the precoat which is held by hydraulic force to a hollow, permeable collector.

The DE filter is a precoat filter (AWWA, 1993) because particulates are removed first on a precoat and then on a cake build-up on that precoat. Because diatomaceous earth has been the predominant media used in treatment, the filter is best known by that media name rather than its functional description.

While DE filtration is strictly a particulate removal process, supplementary treatment may be added to the process train for the reduction of color, organics, iron, and manganese as well as tastes and odors. These constituents are the more common encountered in the average surface water treatment plant operation. Generally, the DE treatment train will be less costly in construction and operation than the more traditional rapid sand filter alternative. The DE filter plant takes much less space for installation and produces a waste that is less problematic for disposal. Because it is practically impossible to cause breakthrough of solids during a filter run, DE filtration may be considered a reduced-risk method of water treatment that requires less operator skills than those needed for the rapid sand filter plant.

Diatomaceous Earth Media

Source

The filtering medium used in diatomaceous earth filtration is composed of fossil-like skeletons of microscopic water plants called diatoms, which range in size from under 5 to more than 100 micrometers. These microscopic algae have the unique capability of extracting silica from water to produce their skeletal structure. When diatoms die, their skeletons settle to form a diatomite deposit.

In the geological past of about 15 million years ago, conditions were especially favorable for diatom reproduction over vast areas like that near Lompoc, California, which was then the ocean floor (McIndoe, Jr., 1969). More than 10,000 species grew there in great profusion in relatively protected waters. The skeletal remains created deposits of almost pure silica—at 1,000 or more feet deep in some location. Eons later, when the land rose from the sea, these marine deposits became available for mining.

Fresh water diatom deposits have also been formed in large lakes or basins shallow enough to permit photosynthesis, having ample nutrients and soluble silica and free of growth inhibiting materials. Such deposits of diatomite are found in many other parts of the world.

Diatomaceous earth or diatomite is distinguished not only by the almost infinite variety of diatom shapes (Figure 1-1) and sizes that make up its composition, but also by its high purity. In its natural state, diatomite is more than 85% SiO_2, which is virtually inert. Most of the remaining constituents are made up of metallic oxides that are also practically inert. The soluble portion of diatomite is extremely low (less than 1%). The odorless, tasteless, and chemically inert characteristics mean that diatomaceous earth may safely be used for filtration of water, beverages, or other liquids intended for human consumption.

FIGURE 1-1. Diatomite Micrograph

Manufacture

Diatomite deposits are mined to be processed by crushing, calcining, air classification, and other methods to produce fine, multi-shaped, angular, porous media. One of the more common diatomaceous media used for drinking water treatment has a mean particle size of 22.3 microns and an average pore size of 7 microns. About 60% of the particle sizes in this media range from 5 microns to 64 microns and have a bulk density or approximately 9 lbs./cu.ft. (dry).

Perlite Alternative

Perlite, having hydraulic characteristics similar to diatomaceous earth and whose bulk density is in the range of 75% of that for diatomaceous earth, is another material used for media in filtration. Perlite is manufactured by crushing and heating a silicious rock to form a mass of glass bubbles. The mass is then milled and air classified to produce media similar to diatomaceous earth, except that the particles are more round than angular.

Application and Historical Background

In addition to the treatment of drinking water, diatomaceous earth filtration today is employed over a broad range of applications including those in the manufacture of foods, beverages, chemicals, and pharmaceuticals. The earliest applications of this type of filtration were in the period immediately following World War I. In the early 1920s, development of diatomaceous earth from the mined diatomite into graded sizes made DE filtration possible in several processes that required a method of removal of particulate matter that would have little effect on filtrate quality. Perhaps the first well-known, major uses of diatomaceous earth filtration, which occurred between 1920 and 1930, were in the recovery of solvents in the dry cleaning industry (Johns-Manville, 1975) and in product filtration in the brewing of beer and refining of cane sugar.

In World War II, the U.S. Army required a new type of filter that was suitable for the rapid, mobile military operations of the time (Black and Spaulding, 1944). An accelerated research and development program resulted in a DE filter that met the basic requirements for a lightweight unit that was easily transportable with the capability of producing drinking water consistent with the military sanitary standards. The DE filtration technology developed during the war was quickly applied to the filtration of swimming pool water and more gradually to the production of drinking water.

Swimming Pools

A somewhat early application of DE filtration was in the filtering of recycled swimming pool water. Some of the more noteworthy early applications occurred in the 1930s, when the motion picture industry "discovered" the effectiveness of this mode of filtration. In filming Aquacades and other underwater ballets of the era, diatomite filtration produced water so clear that bluing had to be added to make it possible to photograph the water.

Both the simplicity and the effectiveness of the DE filter have been important factors in the rapid growth of the swimming pool industry, and have been special influences in the proliferation of backyard pools for individual homes. Homeowners can now operate and maintain the relatively uncomplicated DE units by following the simple instructions provided. A backwashed and cleaned DE unit can be taken out of service in the fall and is ready for operation for the next summer without special start-up measures. The bagged diatomaceous earth medium also stores well. This easy-to-operate process produces pool water as healthful as any alternative mode of filtration designed for this service. With its special ability to remove cysts and other like pathogenic organisms, the DE filter is probably superior for swimming pool application.

Beverage Filtration

One of the first major uses of diatomaceous earth filtration was in the brewing of beer in the 1920s. Applications elsewhere in the industry occurred later. Today, use in the beverage industry is one of the more significant applications of DE filtration. In addition to the clarity of the final product, this method of filtration provides the following two additional benefits critical to beverage production:

1. Diatomaceous earth is inert and will not impart taste, odor, or color to the liquid being filtered.
2. In all DE filters, the precoat cake can be completely disposed of so that all filter runs can be started with a virgin diatomite precoat. Therefore, there is little danger of the growth or recycling of impurities.

The more obvious use for DE filtration would be in the production of clear water that is basic to beverage making. Such water would be essential in the manufacture of soft drinks as well as in the brewing of beer. The more significant applications, however, have been in the integration of diatomaceous earth into the beverage-making process itself. In both the brewing of beer and in winemaking, final separation of unwanted solids is a critical part of the process. This separation is not only essential for clarity of product, but also for the control of fermentation. DE filtration can provide such control and can also produce the clarity desired. In many areas of the brewing and winemaking industry, DE filtration has now become a standard part of the manufacturing process.

What makes beverage filtration especially significant, particularly in breweries, is the contribution to the development of the flat leaf filter of stainless steel construction that is the present mainstay of DE filtration of drinking water. Initially, DE filters employed lined carbon steel shells fitted with tubular elements made of porous material like carborundum. The packed tubular bundle limited the amount of solids that could be removed in a run and inhibited complete sluicing or cleaning before start of the next run. Filtering of beer requires large holding volumes and scrupulously clean septa to avoid contaminating a beer run. These incentives led to rapid development of the present-day flat leaf filter with all its accommodations for efficient sluicing and convenient access to the filter leaves.

Potable Water

The potable water application of DE filtration was literally born of necessity when the United States armed forces required readily transportable water filtration systems suitable for worldwide use in World War II. In general, these filters had to have a high-rate capacity ratio and had to be capable of removing *Entamoeba histolytica* cysts from water. These cysts are common in fresh water sources in most of the hotter, less temperate zones of the world.

The then conventional granular or rapid sand filter did not meet the military's weight and transportability requirements. Although sand filters were adaptable to package assembly in steel tanks, the assembled weight ratio was extremely high and the packed media did not ship well in place. It was often necessary to repack the graded gravel-sand media at each relocation. Moreover, the military sand filters were not effective in the removal of the cyst organisms. Cyst breakthrough was probably due to the lack of pretreatment and operation at excessive flow rates.

In response to the need for immediate action, an extensive research program was initiated at the U.S. Army Engineer Research and Development Laboratories (ERDL). The result of that program was the portable diatomite filter, which effectively met the military's wartime needs. After the war, ERDL conducted a comprehensive program of diatomite filter research to both investigate the basic principles of operation and to develop equipment better adapted to changing requirements.

The earliest municipal diatomaceous earth filter installation was a 75,000 gpd (gallons per day) system used for iron removal in Campbell Hills, Illinois, placed in operation in 1948. By 1977, more than 145 municipal DE plants had been constructed. The estimated total capacity of these units was in the order of 180 mgd (million gallons per day). The largest capacity municipal DE filter installation is the 20 mgd plant at San Gabriel, California. Most of the diatomaceous earth filter installations were made to remove turbidity or particulate matter from water. Fewer installations were needed for water treatment (principally to remove iron and manganese after oxidation by air or by chemical means).

Porous Media Filters

The diatomaceous earth filter is part of the porous media group of liquid filters. Rapid sand filters and slow sand filters comprise the remaining major units used in the treatment of drinking water. The rapid sand filter is named for the original media used in its design. Today, this rapid rate-type filter may use other media such as granular activated carbon (GAC) and granular anthracite. While the unit process features for separation of solids from the liquid flow are the same, significant differences exist between the three major types in configuration, media, production rate, and operation.

The basic intent of drinking water filtration is to separate as much of the particulate solids suspended in the flow-through liquid as possible while inhibiting particles breaking through the media. Breakthrough is the passage of solids due to the inability of the media to intercept such passage or due to failure where the separated solids and media cannot withstand the force of the applied flow. Modern-day filtration of liquids began in the late 1800s with the development of the slow sand filter to treat municipal drinking water. The next development in porous media filtration was the rapid sand filter, which was first used for drinking water treatment around the turn of the century. The diatomaceous earth filter was the last of the porous media filters developed, having been first used for filtration of industrial solvents in the early 1920s and in World War II for drinking water treatment. Each of these three types of filters have similar characteristics and special features that are advantageous for the different applications, discussions of which follow.

Unit Process

The mechanism by which particulate matter is removed in drinking water treatment is somewhat complicated. In many instances, porous media filters will remove particles that are smaller than the openings between media grains. Filtration through porous media depends on a number of physical-chemical processes. The two principal processes or mechanisms by which particles are removed from liquid flows are entrapment and attachment, shown in Figure 1-2.

With entrapment, the filter acts as a sieve or a strainer to physically catch the particles in the liquid flowing through. Suspended matter too small to pass through the available openings are trapped between the grains of filter media. Where entrapment is the more prevalent mechanism of solids separation, most of the particulates will be separated at or just below the surface of the filter media.

The attachment phenomenon has been attributed to Van der Waals' forces or other similar mass or electro-magnetic attraction forces. As the solids bearing liquid is passed through filter media, suspended particulates contact and attach to the surface of individual media grains or to the sur-

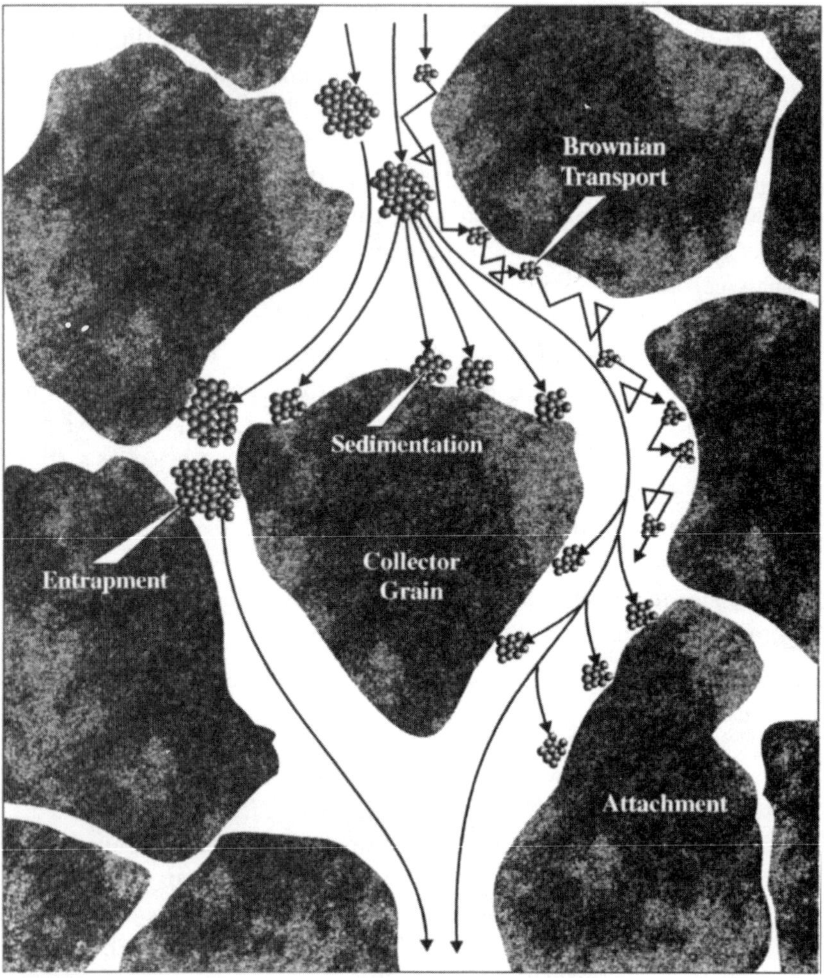

FIGURE 1-2. Principal Mechanism of Porous Filter Separations

face of previously deposited materials. The particulates penetrate the filter media for in-depth separation and storage. For optimal removal of solids by attachment, the particulates must be charge-free. For instance, colloids found in source waters are negatively charged, small-sized particles that tend to hold apart from one another and resist attachment. Coagulation/flocculation, discussed later, will reduce the charge feature and enhance separation by agglomeration of the suspended solids. Agglomeration is the collection of smaller particulates to form larger-sized particles. By increasing the mass, the attachment attraction increases and separation by entrapment is improved.

Progressive Media Change

An important feature of porous media filtration is the change that occurs as the separated solids are accumulated and stored in or on the filter media as the separation process progresses. The stored solids become part of the filter media. Solids separation is influenced by the characteristics of both the filter media and the separated solids. As more solids are accumulated in a filter cycle, the influence of the separated solids in filtration performance becomes progressively more important.

Slow Sand Filter

The slow sand filter uses a somewhat fine grade of sand ranging in size from 0.15 mm to 0.35 mm. Typically, sand is installed about three feet in depth over about 18 inches of gravel. This filter is a gravity type. That is, the surface of the filter is open to atmospheric pressure with about 3 feet depth of water over the surface of the sand. Flow is by gravity through the filter media to collector drains installed under the gravel support bed. The slow sand filter is so called because of its rather low flow rate per unit area of media, which is on the order of 0.05 gpm/sq. ft.

Slow sand filtration may be classified as a "passive" solids removal process. The process is subject to little control by the operator except for adjustment of the filtration flow rate. No chemical addition or solids conditioning is employed. Source water particulate solids are removed by natural physical, chemical, and biological mechanisms. The most important of these are most likely the biological processes that take place on and in the sand media.

Most of the solids are separated in the top few inches of slow sand filter media. This top accumulation has historically been called a "schmutzdecke" (layer of dirt in German). It is sometimes necessary to run a slow sand filter for up to two weeks before the schmutzdecke can form and effective solids removal can begin. The schmutzdecke in the top few inches of media is made up of the separated solids and biological growths that occur in this layer. When the solids and growth build up eventually reduces the head available for flow, the filter is taken out of service for manual cleaning. The top clogged layer is removed hydraulically or by hand, and the filter is then returned to service. It is necessary to replace the depleted sand with new or washed sand depth after a number of such cleanings. Because of the labor intensive cleaning procedure, slow sand filter installations are designed to be cleaned once every three months or more.

Rapid Sand Filters

The rapid sand filter is somewhat similar to the aforementioned slow sand filter in that the design includes use of granular media that is, most often, sand or ground anthracite. The distinguishing features of the rapid sand filter include the following:

1. The ability to intercept and store particulate solids in depth in the media
2. The ability to completely clean and restore the media to its original position by backwashing
3. The ability to operate at media application rates of up to 10 gpm/sq. ft.

The effective size of granular media usually ranges from 0.35 to more than 1.0 mm compared to the finer media used for slow sand filter type. New developments in media design for the rapid sand filter use the coarser grades of media at greater depths of up to 6 feet. Other designs use multiple media where a less dense, coarser media is installed over a more dense, finer grade material. The emphasis on rapid sand filter design is to promote greater depth penetration of the media, thereby increasing the amount of solids removed in a single run between backwashings.

Solids conditioning is mandatory for rapid sand filtration. Some type of coagulation/flocculation is always employed prior to the filter. Coagulation/flocculation provides destabilization or neutralization of charged particles like colloids and the agglomeration of smaller particles to larger sizes to facilitate interception by the media. Soluble organic constituents may also be removed in the coagulation reactions. Effective coagulation is mandatory for successful rapid sand filtration. This filtration process unit, therefore, requires careful control of the operator.

Perhaps the most important advantage of the rapid sand filter is that it may be used for treatment of waters with high-suspended solids content when preceded by a preliminary solids separation device. Coagulation would also be employed to aid such preliminary solids removal. A portion of the neutralized and agglomerated solids can be removed in settling tanks or other solids separation devices in the treatment train before the rapid sand filter. Depending on the type of unit used, up to 90% of the solids can be removed in preliminary treatment. The rapid sand filter can be used in the treatment of water having an extremely broad range of suspended solids and dissolved organic content.

Rapid sand filters may be constructed for gravity flow or may be enclosed in pressure vessels for pressure flow application. Gravity flow is similar to that for the slow sand filter counterpart previously discussed. Underdrain systems, however, are designed to both collect the filtered water downflow and to distribute the reverse backwash flow that is under pressure. The cross-section of the rapid sand filter does not vary significantly from that of the slow-sand gravity alternative except for the underdrain design. In backwashing, pressure applied to the plenum chamber causes the flow to reverse and flow up through the media and thereby wash out the solids separated in filtration. Most rapid sand filters are fitted with supplementary backwash devices like rotating surface wash or air scour to break up mud balls and other agglomerated media. The rapid sand filter configuration may also be enclosed in a tank for pressure operation.

Diatomaceous Earth Filters

The diatomaceous earth filter differs profoundly in arrangement, but not in its particulate removal functions, from the two granular media filters previously discussed. The media used is powder-like in characteristic, with individual grain sizes predominantly ranging between 5 and 25 microns. The solids separation phenomenon is similar to that of the slow sand filter as:

1. Separation occurs predominantly in the surface of the media.
2. The surface media may be predominantly the separated particulates or a mixture of such particulates and supplemental media.

In flow application, the diatomaceous earth filter is closer to the rapid sand design. Application rates for drinking water treatment may range from 0.5 to more than 3.0 gpm/sq. ft., much higher than the average rate for slow sand filters.

In DE filtration, a thin protective layer of diatomaceous earth is first built up, or accumulated, on a porous filter septum (a permeable cover over interior collection channels) or membrane. The septum is most often plastic or metallic cloth mounted on a wire mesh covered steel frame.

Sometimes carborundum tubes have been used where the inherent permeability obviates the need for a cloth covering. The DE process is also called precoat filtration because the solids separation at the start of a run takes place on the built-up precoat layer of diatomaceous earth.

The various steps in the precoat filtration process are shown in Figure 1-3. Precoat is applied by recycling a diatomite slurry from a precoating system, through the filter, and back to the precoating system. This recycle continues until the entire septum is covered with a layer of from about 0.10 to 0.20 lbs/sq. ft. of filter area with a reasonably uniform thickness. As shown graphically in Figure 1-3, the sizes of individual media particles are usually smaller than the openings in the woven cloth septa. The precoat is actually retained by bridging or arching of the particles across the openings. It is this feature that makes it important that no unusual change in pressure occur across the septum during filtration. It is also essential that no other forces be brought to bear on the precoat like the passage or discharge of gas bubbles. When the precoat operation is complete, filtration is initiated. Particulate solids in the product flow are separated on the precoat surface. With such separation, the unwanted particulate matter actually becomes part of the filter media. If the characteristics of the separated matter are such that reasonable porosity is maintained and the surface is not clogged or blinded, filtration continues until a maximum head loss across the septum is reached. At such time, filtration is halted and the filter cake is removed.

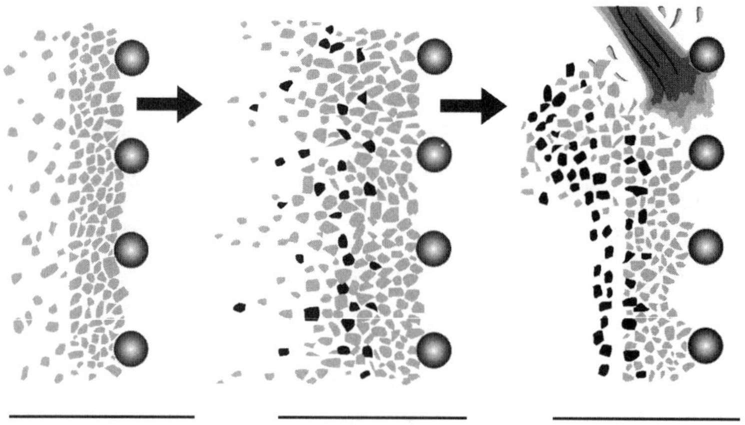

Precoat Formation Filter Cake Formation Filter Cake Removal

FIGURE 1-3. Precoat Filtration Operating Cycles

In most filters, the cake is removed in slurry form by sluicing. High-pressure sprays, directed at the accumulated cake, detach the cake and provide dilution for draining the slurry suspension from the filter vessel. When cleaned, the filtration operation is repeated, beginning with the precoat cycle.

Many times the suspended matter being separated in filtration will have features that tend to blind the filter media. Such material might be slime-like in character, algae that is filamentous in shape, or simply suspended matter with a large proportion of fine particulates. Other separated solids may be compressible in nature. With separation, these solids have the capability of matting or plugging the precoat to the extent that much of its porosity or permeability would be reduced. To overcome this effect, an additional filtration feature is employed. A predetermined amount of diatomaceous earth (called "body feed") is added to the incoming flow to provide a matrix for the accumulated separated solids. This DE matrix holds the separated particulate matter apart to increase the overall porosity of the cake. With proper proportioning of body feed, blinding of the precoat is inhibited, head loss is reduced, and the length of filter run is extended.

The body feed proportion required to maintain a filter cake of uniform porosity will vary with the concentration and nature of the solids to be removed. Uniform cake porosity is that porosity that will produce the lowest headloss for the production of a specific volume of filtered liquid product of acceptable clarity. Other variations in diatomaceous earth filtration have been developed for special filtration needs. These variations are discussed in more detail later.

The Diatomaceous Earth Filter and Pertinent Regulations

General

The effectiveness of a DE filter in drinking water treatment would first be determined by the ability to produce finished water quality meeting the standards of the pertinent drinking water regulations. The Safe Drinking Water Act, initially passed in 1974, and its subsequent amendments provide the basic rules and limitations regarding water quality. Where individual states have accepted primacy, the rules, which can be extremely stringent, are enforced by the state rather than the federal EPA. In a complex development process, involving both difficult scientific issues and political considerations, the USEPA is continually adjusting, adding to, and expanding the scope of the established regulations. Because of this continuous change, the design engineer must be aware of proposed rules and rules under development to avoid the possibility of filtration obsolescence or, at the least, the need for costly alterations when new rules are finally established.

The current established EPA drinking water quality regulations include long lists of maximum contaminant levels for inorganic and organic constituents as well as those rules and limits related more directly to the filtration process train. The discussions that follow are limited to those rules directly related to filtration or to those for treatment features that may be added to a filter train to provide multiple benefits.

The primary function of filtration in drinking water treatment is the removal of particulate matter. As specified in the regulations, the effectiveness of filtration is evaluated as the reduction in turbidity (inlet to effluent water). Turbidity, a measure of light scatter in a batch or continuously running sample, is considered to be a surrogate for the measurement of particle content. More recently, particle count equipment measuring particles above 2 microns in size, are being used more frequently.

Table 1.1 lists the principal water quality parameters in the current regulations that serve to monitor the overall effectiveness of a filter train treating potable water. As indicated, these parameters are divided into primary and secondary groups. The primary regulations address health problems and may be considered to be absolute in the stipulated limitations. The secondary regulations are related to aesthetic issues where some leeway may be permitted. Some secondary issues, such as iron and manganese limits, are being considered for inclusion in the primary listing. Moreover, the average water system customer knows what he sees so that, whether required or not, a water system will produce water of good color or iron and manganese features.

Each of the water quality issues in Table 1.1 are discussed in reference to the regulatory limits and trends and treatment that may be provided by porous filtration process trains in general and diatomaceous earth in particular.

Turbidity

As noted in Table 1.1, turbidity, usually measured in nephelometric turbidity units (NTU), is a measure of light scatter that indirectly determines

TABLE 1.1. Principal Water Quality Parameters and the Method of Measure

Parameter	Unit	Description
Primary SDWA Regulations		
Turbidity	NTU	A measure of light scatter through water under certain conditions. Indirectly, a measure of the concentration of suspended particulate matter in water.
Pathogens	Count	Measure of the content of coliforms and the cysts of *Giardia* and *Cryptosoridium* by microbiological assay methods.
TTHM	mg/l	A measurement of the total concentration of trihalomethanes in a water sample, as measured by the standard spectrophotometric methods. TTHM concentration is composed of the sum of concentrations of the four most prevalent trihalomethanes: chloroform, bromoform, dichlorobromomethane and chlorodibromomethane.
Secondary		
Color	scu	A measure of light absorbance by a given water sample at a given light frequency, used as a measure of consumer acceptability. Indirectly, a surrogate measure of the concentration of organic matter in water.
Iron Manganese	mg/l mg/l	A measure of the total concentration in a water sample, as measured by the standard wet chemical methods. Important for consumer acceptance, particularly in terms of staining of fixtures and clothing.

the suspended particulate content. Turbidity has been used for decades as the standard in determining the effectiveness of particulate or suspended matter removal. With the development of reasonably accurate particle counters, however, the limitations of the turbidity measurement has been disclosed. Sensitive to the type and form as well as the concentration of particles, a higher turbidity measurement of a source water at one time of the year may actually have less particle matter than at another time when turbidity is indicated to be lower. Tests in City of New York research have shown that, in parallel treatment of the same water, the particle count of the DE filter effluent may be less than half of that for rapid sand filters when the turbidity readings are similar. Use of particle count rather than turbidity, therefore, should show the DE filtration product to be lower in suspended matter than the alternative porous filter methods. The EPA turbidity standards recognize this difference and has established the following finished water maximum contaminant level (MCL):

— Rapid Sand Filters: 0.5 NTU
— Diatomaceous Earth Filters: 1.0 NTU

While suspended matter is the measure, a primary intent of solids reduction is the reduction of the pathogen content. Bacteria and viruses may be found in drinking water sources attached to particles. The cysts of *Giardia* and *Cryptosporidium* can be discrete particles between 2 and 5 microns in size. Current particle count equipment can ascertain sizes 2 microns and larger. Turbidity or particle reduction in filtration, therefore, is directly related to the control of pathogens shown inTable 1.2.

In an extensive test program of porous filters that provided the basis for the above federal turbidity limits, it was demonstrated that the DE filter also had superior cyst removal ability. Background data from this and other programs were used to develop the disinfection credits for baseline filtration shown in Table 1.2. Pathogen removal and inactivation is considered to be a stepped process. The more pathogens removed by filtration, the less that must be inactivated by subsequent chlorine or other

TABLE 1.2. Baseline Filtration: Disinfection Credits

	Maximum Turbidity[1]	*Giardia lamblia* Removals		Virus Removals	
Filter Process	NTU	Log	%	Log	%
Conventional: coagulation, flocculation, sedimentation, rapid rate (granular) filtration	0.50	2.5	99.3	2.0	99.0
Direct filtration: coagulation, rapid rate (granular) filtration	0.50	2.0	99.0	1.0	90.0
Slow sand filtration	1.0	2.0	99.0	2.0	99.9
Diatomaceous earth filtration	1.0	2.0	99.0	1.0	90.0
Total (including disinfection)[2]	—	3.0	99.9	4.0	99.9

[1]Maximum turbidity limits must be achieved in at least 95 percent of the measurements made. Turbidity of drinking water must never exceed 5 NTU.

[2]If the presence of *Giardia* cysts is estimated to exceed 1/100 1 (one cyst per 100 liters), greater inactivation is required.

type of disinfection; hence, the higher credits for DE filters for *Giardia* cyst removals. Lower virus credits are allowed for DE. This may be assumed to be influenced more by the one-step solids removal as compared to the two-step conventional rapid sand alternative and the fact that virus may involve much smaller particles. Moreover, the rapid sand filter process uses coagulation to neutralize charged particles and agglomerate smaller solids into larger-sized floc for improved removal.

While DE filtration involves strictly physical separation, supplemental treatment may be provided to improve the removal of smaller particles, especially those of colloid size. Despite the small pores, measured as fractional microns, small, colloidal charged particles like negatively charged natural organic matter (NOM) can pass through a filter cake. If the charge is neutralized, colloids are separated by attraction in the DE filter cake, just like in the large media rapid sand filter. Since most charged particles are organic in nature, preozonation may effectively break the bond between the organic and inert matter, thereby neutralizing the charge. In another method, part of the DE filter precoat media may be coated with positively charged alum. The negatively charged colloids are then attracted to the positively charged media.

Because of increased concern over the proliferation of pathogenic cysts, especially those of *Cryptosporidium*, there is no doubt that future regulations will call for improved solids removal and reduced solids content in limited drinking water. The first step is expected to be a turbidity limitation of 0.2 NTU. Many current research programs have established a goal of less than 0.1 NTU in anticipation of the more stringent future standards. In any event, filter comparison tests have demonstrated lower particle count for DE filter effluent for similar turbidities. It is, therefore, possible that the EPA standards may continue the higher turbidity allowances for the DE filter alternative.

The use of DE filters is limited to treating source waters with the high range of turbidity between 5 and 10 NTU. Where turbidity is caused by inert silt-like particles, the higher range may be considered. Where turbidity is mostly organic in nature, 5 NTU would be the probable limit.

Disinfection Byproducts

Trihalomethanes, suspected of being possibly carcinogenic and/or mutagenic to humans, are the only disinfection byproducts (DPBs) listed in the federal drinking water regulations as of 1996. As indicated in Table

1.1, the maximum contaminant level (MCL) is actually the sum of the content of the four most common trihalomethanes (THMs).

Trihalomethanes are generated in water treatment through the reaction of natural organic precursors found in many source waters and chlorine used in disinfection. Of the byproducts included in the regulations, the more prevalent of these include a group of haloacetiles, also formed by chlorine reaction, in addition to aldehydes and ketones formed through the reaction with ozone, another disinfectant gaining popularity in drinking water treatment. The expanded list is shown inTable 1.3.

The EPA is considering changes to the DBPs list to include more stringent MCLs for THM and HAA5 (the sum of 5 specified haloacetiles). These proposed changes are as follows:

THM
– Current MCL: 80 microns
– Ultimate MCL: 40 microns

HAA5
– Current MCL: 60 microns
– Ultimate MCL: 30 microns

TABLE 1.3. Principal Disinfectant Byproducts: Disinfection/Disinfection Byproduct Rule

Disinfection Byproducts	MCL (mg/l)
Total trihalomethanes (TTHM: sum of 4)	0.080
Chloroform	
Dibromodichloromethane	
Bromochloromethane	
Bromoform	
Haloacetic acids (HAA5: sum of 5)	0.060
Dichloroacetic acid	
Trichloroacetic acid	
Monochloroacetic acid	
Dibromoacetic acid	
Monobromoacetic acid	
Chlorite	1.0
Bromate	0.010

Disinfection byproducts are particularly significant when considering the use of DE filtration for water treatment. Unlike the other porous filter methods that employ chemical coagulation, DE filtration will not remove the dissolved organic precursors. Ozonation can be an important pretreatment in the DE filter train; chlorination has almost universal acceptance both for disinfection and for maintaining a residual in water distribution systems.

In rapid sand filtration, coagulation with metallic salts will remove a significant proportion of organic precursors, especially those of high molecular weight. A small percentage of the lower molecular weight, more biodegradable organics may be removed by biological action in the granular media. Up to 25% of precursors may be removed by biological action in slow sand filters because of the long detention time, up to 10% in rapid sand filters.

Since most surface waters contain some natural organic matters, it is essential that the potential for DBP formation be investigated thoroughly when considering the use of DE filters. Where the precursors are low, alternative disinfection regimes beyond the scope of this text may be used effectively to limit DBP formation.

With the trend toward more stringent DBP requirements, it would appear that biologically activated carbon media (BAC) may need to be adopted for many surface water applications. In addition to reducing the level of DBPs, it will be important also to reduce the level of biodegradable dissolved organic carbon (BDOC) in filter effluents. Any strong oxidant like ozone or chlorine will convert high molecular weight organics to the more assimilable low molecular weight BDOC form. Any loss of chlorine residual in a distribution system may lead to regrowth of bacteria protected in pipeline sediment or coating. Where such potential exists, it may become necessary to add BAC to or integrate BAC into the filter train. In the case of DE filtration, providing preozonation and BAC units before the DE filters has been found to be most effective. With this arrangement, discussed in more detail later, both organic precursor and BDOC levels are reduced. Ozone provides the multiple benefits of solids conditioning, inactivation of pathogens, and conversion of high molecular weight organics to enhance BAC and DE filter performance in the production of a safe drinking water supply.

Color

Color, tastes, and odors, when present in drinking water, are probably the most distinctive features that disturb the typical consumer. While for the most part not health threatening, it is the aesthetic values of water that are first noticed. More often it is the natural organic content that influences the presence and extent of color, tastes, and odors in source waters. In less frequent instances, industrial wastes may be the cause.

Color may also be caused by iron and/or manganese that are present in either colloidal or dissolved form or both. With coagulation preceding rapid sand filters, the charges of both organic and metal colloids may be neutralized and then agglomerated into the hydroxide floc. Dissolved organics and metals are adsorbed by the floc so that coagulation alone may be sufficient to reduce source water color to below 5.0 color units (CU), the established acceptable goal.

Partial color removal may be accomplished in DE filtration. Treatment of a northeast U.S. reservoir supply with raw water color ranging from 30 to 60 NTU with prechlorination reduced color to below 10.0 CU at all times. It was necessary to use ozone, a much stronger oxidant, to maintain color below 5.0 CU continually. Ozone is also more effective in breaking the bond between organics and inert colloids to neutralize negatively charged NOM. Ozonation and other oxidants will precipitate and otherwise condition dissolved iron and manganese for removal in the DE cake. A discussion about iron and manganese removal follows.

A significant part of color removal with strong oxidants like chlorine and ozone is the bleaching action. These oxidants have been used for years in industry as well as drinking water treatment for color reduction. Because of the DBP formation potential, it is essential that both the effectiveness of oxidant in color removal and the possible impact on generating unacceptable DBP levels be carefully studied for the particular source water before selection of treatment.

Iron and Manganese

As previously mentioned, the first step in any iron or manganese removal process would be oxidation to generate a precipitation action before removal in the filter media. Oxidation may be provided by chemical oxidants like magnesium oxide (MgO_3) for iron and potassium permanganate for manganese. Ozonation and chlorination may provide alternate methods for oxidation.

In rapid sand filtration, iron precipitates are agglomerated in flocculation and removed by the filter media. For the most part, oxidation of manganese does not form discrete particles. The manganese is rather conditioned for removal by "scaling" any surface contacted. The granular media in rapid sand filters becomes coated with a soft manganese scale, eventually reaching a point where the turbulence of backwashing will remove the outer coating. In fact, once established on the media, the manganese coating catalyses the scaling action for improved removals. A significant advantage of iron and manganese removal in rapid sand filtration is the volume of media fixed in place to accept the separated solids and the time of travel through the media for that separation. For instance, at a filter application rate of 6 gpm/sq. ft., time of detention in media 30 inches deep is about three minutes. The pore volume used to hold the separated iron and manganese oxides would be a fraction of that needed for removal of the other flocculated suspended matter.

By nature of design, DE filters do not afford significant time of detention in the media to allow formation of precipitates or scale. The precoat is generally 1/8 inch thick, building up to between ½ and ¾ inches in total depth. At 2 gpm/sq. ft., DE filter application rate, detention through a ¾-inch cake would be about 5 seconds, an insignificant length of time to permit scaling of manganese. On the other hand, the fine powder-like DE media provides more than enough surface area, but not the time for the required action to happen. In a DE filtration process train, this deficiency factor is overcome by installing a detention tank for the chemical reactions to occur before the filter. Discussed in more detail later, a detention basin providing about 10 minutes time will allow either the iron precipitation to occur or permit manganese to scale on the body feed added concurrently with the oxidant. Depending on the magnitude of content of each metal, a single detention basin may be sufficient, but high iron content may require two separate steps of conditioning, detention, and filtration. Again, the DE filter acts simply to remove particulate matter either preformed or scaling the body feed. As will be discussed later, it may be possible to retain in detention or to recycle manganese-coated DE media for extended use and more efficient removal. Again, pilot research would be essential to determine the best method for iron and manganese removal.

Tastes and Odors

Tastes and odors, a secondary quality issue, are removed from drinking water principally through oxidation. The same aeration, chemical oxidation (potassium permanganate) and gaseous oxidation (ozonation and chlorination) that would be used in the rapid sand filter train for other purposes would precede the DE filter. Some oxidants, particularly chlorine, may exacerbate some taste and odor problems so that pilot research would be necessary to determine the best method for a particular location.

Operating Efficiency

Previous discussions were related to water quality issues. In considering the method of filtration to adopt for a particular application, the next most important feature would be efficiency of operation as it may affect labor and overall costs. All porous filters involve batch operation. That is, any filter may be operated until the solids buildup in the media inhibits flow sufficiently to require cleaning to restore its hydraulic capability. In the case of the rapid sand filters, the separated solids clog the media interstices and are removed by backwashing. In DE filtration, the buildup of separated solids in the separated cake gradually increases hydraulic head loss through the filter until the cake must be sluiced and the filter cycle (precoat, etc.) can be started again. The longer a filter run may be extended, the less the amount of labor effort or sluicing water costs would be expended.

In DE filtration, not only the amount of solids to be removed, but the character of the solids will affect the efficiency of removal. Because a somewhat high hydraulic force must be applied to hold cake on the filter leaf, compressibility of separated solids becomes an important factor. If these solids are predominantly silt, little compression will occur. If, on the other hand, the solids are essentially biologic or plant detritus, the cake can compress, which reduces the available flow through pore size and the length of run.

In a similar manner, size of separated solids will affect pore size. With a broad range of particle sizes, the smaller solids tend to "nest" in the interstices to reduce the size of the flow path and accelerate loss of head before cleaning must be initiated. Both the size and compressibility problems in DE filtration are mitigated by conditioning the cake through the addition of the body feed to the inlet flow of the same DE material used for precoat of the filter.

By adding body feed to the inlet flow to combine with the source water solids, body feed particles provide a matrix surrounding the potentially troublesome raw water matter. Body feed literally holds the troublesome solids apart to effectively maintain the porosity of the cake. With sufficient body feed, the head loss buildup should be more a function of cake thickness than decreasing porosity. Depending on the character of the raw water, body feed may have to be added to the incoming flow in proportions many times that of the source water solids by weight. Depending on the specific application, it would not be unusual for body feed requirements to be as high as 10 to 50 times (by weight) the particulate solids carried in the inlet flow. Selection of optimal body feed rates is discussed later in the text.

Multiple Design/Operation Options

It is apparent from the description of DE filter train options that several individual treatment additions provide multiple benefits. In the design of a DE filter train, it is essential that these alternatives be considered to ensure both the most efficient provisions for current water quality needs as well as changing future regulations.

A number of treatment building blocks are used to develop a DE filter train and expand its capabilities. While the optional blocks may be somewhat limited in number, the various combinations do provide for handling the most common treatment problems encountered in drinking water treatment. As indicated in Table 1.4, starting with basic DE filtration, multiple benefits may be added in three additional stages. As indicated, the additions all involve pretreatment expanding the system as follows:

— Step 1: Basic DE filtration
— Step 2: Add chemical oxidation and detention
— Step 3: Add preozonation (detention other than ozone contactors not required)
— Step 4: Add BAC (biologically activated carbon media) contactors

Each of the pretreatment building blocks will involve additional, but not significant loss of head before the filter. The additional steps involved may be compensated by increasing the TDH (total dynamic head) of the filter pump. If the building of the DE train is expected to occur over an extended period of time, sufficient site area must also be provided for the additions. Considering the rate and scope of changing regulations, it would be advisable to provide both additional hydraulic head and site area for changes not yet anticipated.

A discussion of the salient features of each of the building steps follows.

TABLE 1.4. Summary of Principal Multiple DE Treatment Benefits

1. DE Filter-Particulate Removal.
2. Chemical Oxidation, Detention + DE Filter
   - Iron and manganese reduction
   - Reduction in chlorine demand and the level of THM formation
   - Partial reduction of color, tastes, and odors.
3. Ozonation + DE Filter
   - Reduction in slime-like quality of organic particulates to reduce compressibility and extending length of filter runs
   - Neutralization of negatively-charged NOM
   - Significant reduction in chlorine reaction, sites, and the level of THM and other DBP formation
   - Inactivation of pathogens
   - Reduction in color, tastes, and odors
   - With 10 minutes detention—reduction of iron and manganese.
4. Ozonation + BAC Contactor + DE Filter
   - All of 3.0 above
   - Conversion of high molecular weight organics to assimilable BDOC removed by BAC-reduction in DBP formation
   - Oxidation of iron and manganese for principal removal in BAC media
   - Partial reduction of source water solids to reduce solid removal load on DE filter for extension of length of run.

Basic DE Filtration

As noted in Table 1.4, the basic filtration train simply removes particulate matter from source waters. As reviewed in a later chapter, straight DE filtration may also remove a limited amount of color without supplemental treatment. The basic filter train would be sufficient for source waters of:

— Low to moderate turbidity
— Low color
— Low organic content with low trihalomethane formation potential.

An ideal, economic application of DE filtration would be for well supplies under the influence of surface waters, but otherwise acceptable in quality. The superior cyst removal capability would make the DE filter more advantageous than other alternatives.

Chemical Oxidation and Detention

This addition, Step 2 in Table 1.4, requires a basin or basins to provide at least 10 minutes detention at rated flow and metering pumps to add either magnesite (MgO) for iron removal or potassium permanganate ($KMnO_4$) for manganese removal. Ozonation might be a much more dynamic oxidant for both iron and manganese, but the more sophisticated equipment and safety precautions required would make chemical oxidation a more practical application for the small capacity plant or plant with limited staff capability.

If both iron and manganese must be removed and a significant amount of iron is present, removal may have to be accomplished in two successive, complete trains with iron removal preceding manganese. If iron is low enough in content, removal of both constituents may be possible with just the permanganate oxidation.

As noted in Table 1.4, the additional benefits of reduction in chlorine demand, THM formation, and tastes and odors may be realized with chemical oxidation. It is emphasized that the extent of these additional benefits depends on the quality features of the source waters.

Preozonation

Ozone is the most dynamic of all available oxidants. Use of preozonation will significantly extend the treatment capability of the DE filter train. It is especially effective where the particulate matter in source waters is predominately organic in nature. Where high organic content exists, the more obvious benefits of preozonation would be neutralization of colloid charges for improved filtration removal, reduction in tastes and odor levels and the reduction in DBP formation potential. In a relatively short time of contact, ozone will inactivate the most resistant of pathogens including those of *Cryptosporidum.*

The ozone benefit most pertinent to DE filtration is that of reducing the compressibility of the suspended matter in source waters. Particulate matter in even moderately eutrophic reservoirs and lakes will be biologic in nature or may be inert matter covered with a biologic growth. Slime-like biologic or biologic-covered particles separated in the DE filter cake will typically compress due to the applied hydraulic pressure. When the cake compresses, its porosity is reduced, which increases the rate of head loss buildup and shortens the length of run. The increase in length of run resulting from preozonation can be significant. Ozone dissolves and breaks the bond of organic matter with raw water colloidal-sized particles. The negative charge of colloidal particles is thereby neutralized and overall compressibility reduced.

While ozonation does provide significant advantages in the treatment of water with particulate organic content, there is another feature that would cause problems. Ozone will convert high molecular weight dissolved organics to low molecular weight, more biodegradable material. With sufficient BDOC in the product water, regrowth problems might be enhanced in the water distribution system. With conditions of insufficient chlorine residual in the water, organisms shielded by pipeline sediment or pipeline growths, that may be pathogenic, could multiply to possibly cause infections. For this reason, the organic quality generated by ozonation should be investigated to determine whether an additional biologically activated carbon (BAC) contactors may be required to reduce the BDOC and produce a reasonably stable water.

Biologically Activited Carbon Contactor Addition

The BAC contactor is a column or basin filled with granular activated carbon (GAC) to furnish a large surface area on which to promote biologic growth.

Ozonation converts the less assimilable organics to the biodegradable form. The total BDOC provides the food for the biologic growth. With sufficient time, the BDOC should be reduced sufficiently to produce a stable water to inhibit the regrowth potential and to further reduce the DBP formation potential. Ozonation will also oxidize iron and manganese for removal both in the BAC contactors and in the final DE filter.

An added benefit of this ozone-BAC-DE filter train is the partial removal of particulate matter in the contactor to extend filter runs. With BAC preceding the DE filter, another advantage over alternative methods is offered. Where the filter media in rapid sand filters is used for the BAC action or in another common arrangement where BAC follows the filter, any expended biologic growth particles that may slough off in the contactor is discharged with the product water. In the DE filter train, the filter catches and retains such biologic particles in the cake.

Non-Structural Options

There are other non-structural operation alternatives that may be adopted to improve the quality of DE filter output. Several grades of diatomite are available to provide reduced porosity in the precoat and possibly in the cake through body feed. This means that if finer particulates must be removed because of seasonal changes in source water quality, optional grades of DE are available.

It is also possible to alum coat part of the precoat to enhance removal of colloids. This may be used instead of ozonation to neutralize colloidal charges. With alum coated precoat, the positive charged aluminum will attract and hold the negatively charged colloids in the natural organic matter (NOM). This would provide a viable option for the more complicated ozonation alternative, especially for small systems.

Preliminary Research

It is essential that both benchtop and pilot scale research be conducted to determine the viability of and the design criteria for DE filtration systems. No two application situations will be the same. Research procedures are outlined in a later chapter. All investigations should take the possibility of changing drinking water regulations into consideration. Pilot tests will not only provide the basis for the most economical design, but more importantly will avoid costly mistakes in regard to design for both current and future conditions of source water quality and drinking water regulations.

2
Diatomaceous Earth Filter Types and Options

Unlike the rapid sand filter where most installations are similar in appearance and hydraulic accommodations, the DE filter can have many variations in design. The basic filtration function, however, is always the same. Today, practically all new DE filters are fitted with flat leaves covered with a septum or woven cloth membrane to support first a precoat and then cake of separated solids as the filter cycle proceeds. The principal differences in DE filter design are in the hydraulic arrangements for introducing and discharging flow and in the method of accessing filter internals.

General

Hydraulic Groupings

DE filters must be separated into two distinct groupings, defined by the method of introducing flow and providing the hydraulic force to hold the cake to the septum. All flows are controlled and maintained by pumping in both of the following arrangements:

— *Pressure.* Flow is introduced to the inlet of the filter and is controlled by monitoring the discharge rate. The filter elements are contained in a pressure vessel.

— *Vacuum.* The pump suction is connected to the effluent manifold for the filter leaves. The top of the filter is open to the atmosphere and the flow rate is monitored at the pump discharge. Inlet flow is admitted to the open filter vessel by means of float control to maintain a constant level.

Vacuum filters were introduced later as a more simple, economical method for filtering water at lower capacities. Both types provide the same efficiency of solids removal.

Principal Elements of Flat Leaf Filter Design

Figure 2-1 illustrates the principal elements that would be common in the manufacture of any flat leaf filter used in the treatment of drinking water. As shown, the principal elements of all DE filters include the following:

— Containment vessel
— Baffled inlet
— Filter leaves mounted on an effluent manifold
— A method or system for sluicing the cake from the filter leaves at the end of the run
— A drain to receive the sluiced slurry
— An open top or other access mode.

The principal variations in the design of these common filter elements would be in regard to:

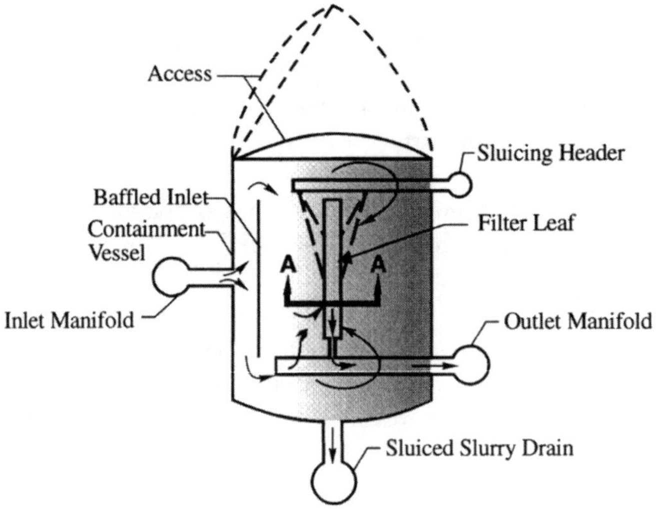

FIGURE 2-1. Principal Elements—Flat Leaf Filter

— The configuration, assembly, and installation connection of filter leaves
— The configuration and mounting arrangement of the containment vessel, mode of access, and flow inlet to vessel
— The manner of sluicing or cleaning cake from the filter leaves.

Filter designs also vary with regard to the type of materials used in the fabrication of the component parts and in the method of accessing the filter vessel for inspection and maintenance of internals.

Manufacturers provide several arrangements of the elements shown in Figure 2-1 to accommodate particular filter applications. These accommodations may be for size and capacity, economy of installation and operation, in addition to treatment requirements. Because of the many options available, it is possible to consider a number of types of filter designs and arrangements to determine "best fit" for both installation and improved operating efficiency. The primary intent in this chapter is to familiarize the design engineer with the different types of DE filter equipment and equipment options available from various manufacturers to provide guidelines for meaningful equipment selection.

Initial Filter Designs

The practice of diatomaceous earth filtration for drinking water is relatively new compared to the century-old slow sand and rapid sand alternatives. The earliest DE filter units were fabricated for use by the U.S. Army in the field in the early 1940s. Operating under pressure, these early units were basically vertical tanks fitted with a plate or tube sheet at the top from which cylindrical septum elements were suspended. The filtrate collection chamber was at the top; flow entered the bottom. In the first designs, cylinders of porous refractory materials or helically wound wire were used.

Figure 2-2 is a cross section of a typical 50 gpm diatomaceous earth filter unit used by the U.S. Army in World War II. After precoat, inlet water enters the bottom of the shell where entrance velocities are dissipated. During filtration, water flows in an upward direction to deposit solids on the cylindrical septa. At the end of the run, flow is reversed by entering the top of the shell and then flowing through the cylinders to remove the accumulated cake. The washed slurry flow exits the unit through the bottom conical sections. The unit shown has a DE slurry

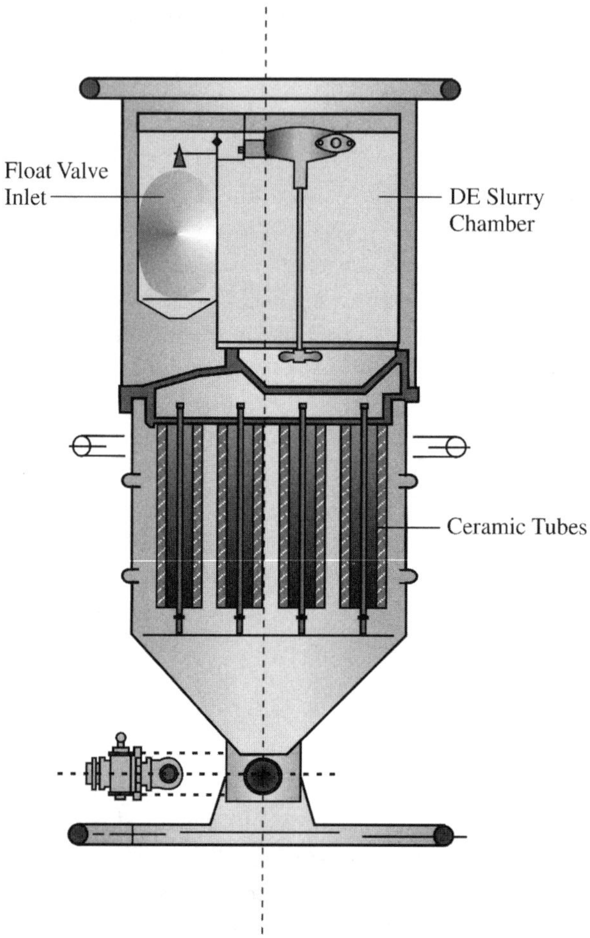

FIGURE 2-2. Section Early Military Tubular Element Filters

tank with mixer and suction float control box mounted on top of the filter shell. The float box controls the entrance of DE slurry for precoat and body feed to the filter pump suction (not shown). This relatively lightweight 50 gpm unit supplanted the previous standard module pressure sand filter units that weighed almost 5,000 lbs mounted on skids.

Figure 2-3 shows a later development of the tubular design fabricated for commercial use. These later designs were larger capacity and more efficient than the original units. The basic difference in the new units was the longer cylindrical septum elements installed closer together to provide greater septa area for unit volume of containment.

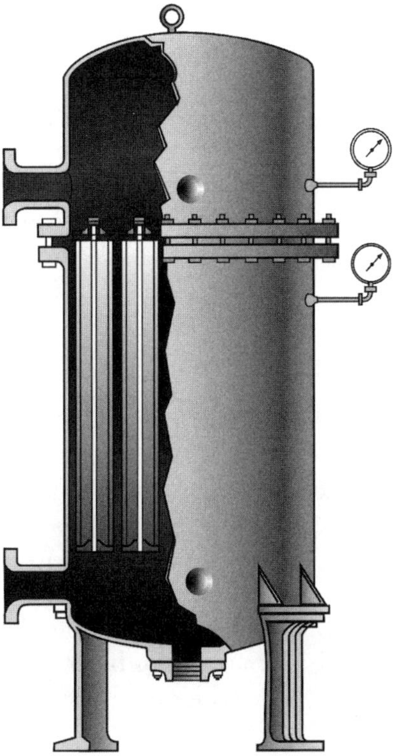

FIGURE 2-3. Later Tubular Element Filter

In terms of septa area, the cylindrical element filter is a very efficient design. The somewhat closely packed tubes provide significantly more surface area per unit of tank volume than the present-day flat leaf design discussed below. There are, however, inherent problems in the operation of tubular DE filters as follows:

— There is a tendency for bridging where the accumulating cake on adjacent tubes eventually touch to block even distribution of flow.
— Backwashing or even sluicing may not provide uniform cleaning, and portions of the tubes will clog, especially with porous ceramic or carborundum. This can lead ultimately to uneven or incomplete precoating.
— There is no easy access to the tubes for supplemental cleaning or for maintenance. The top of the tank and the tube sheet with tube

assembly must be lifted out for such access. Even with removal of the tube bundle, it is very difficult to inspect or to work on the surfaces of interior tubes.

In spite of these problems, the tubular filters have provided invaluable service in the filtering of drinking water. Filters of this design are still in service in many plants in the United States. Most new designs of DE filters used for drinking water have vertically mounted, flat leaves as discussed later.

Vertically Mounted, Flat Leaf Filters

The vertical, flat leaf filter design evolved as a result of two motivating influences. First, the development of stainless steel wire and plastic filament cloths mounted on rigid flat frames provided a leaf that reduced the more significant problems of the tubular filter. Second, the use of DE filters in many parts of the beverage industry generated new designs using stainless steel and plastic for fabrication of many of the structural filter elements. In the brewing of beer, DE filtration is much more common than in drinking water treatment. This beverage industry use resulted in designs that not only provided improved operation and performance, but also the development of much larger capacity filter units than was possible with the tubular design. With individual units of up to 10 mgd capacity, it is now possible to consider the use of DE filters for large as well as small capacity drinking water treatment facilities.

Figure 2-4 shows the general arrangement of a flat leaf filter with a vertical containment similar to the tubular unit, Figure 2-3. The flat, vertically mounted filter leaf provides many advantages, the principal of which are as follows:

- — The accumulated cake can be sluiced off the leaf by means of high-pressure sprays directed at the sides of the leaves. This provides for much more efficient cleaning. For filters with easy access, the cake may be hosed off manually.
- — Individual leaves, provided with nozzle outlets designed for convenient connection to an outlet manifold, may be easily removed for replacement and/or repair.

The new stainless steel and plastic flat leaf designs allow consideration of DE filtration for a much broader range of drinking water treatment appli-

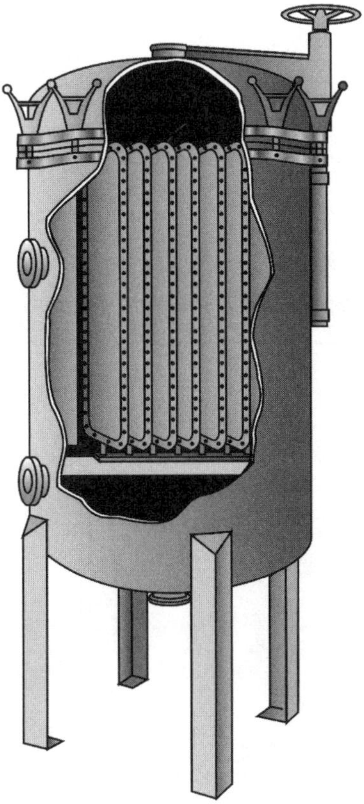

FIGURE 2-4. Vertical Flat Leaf Filter

cations. Now the principal features limiting the use of DE filtration are more related to source water quality and certain process requirements.

Hydraulic Modes of Operation

The two basic types or groupings of diatomaceous earth filter designs that are essentially defined by the hydraulic mode of operation are shown on Figure 2-5.

Both the pressure and the vacuum type of filters remove particulate matter equally as well. The principal differences are related more to the filter leaf application rate of flow and the length of filter run that would be permitted by the differential head available in the two modes of operation. These and other salient features of the two types are discussed later.

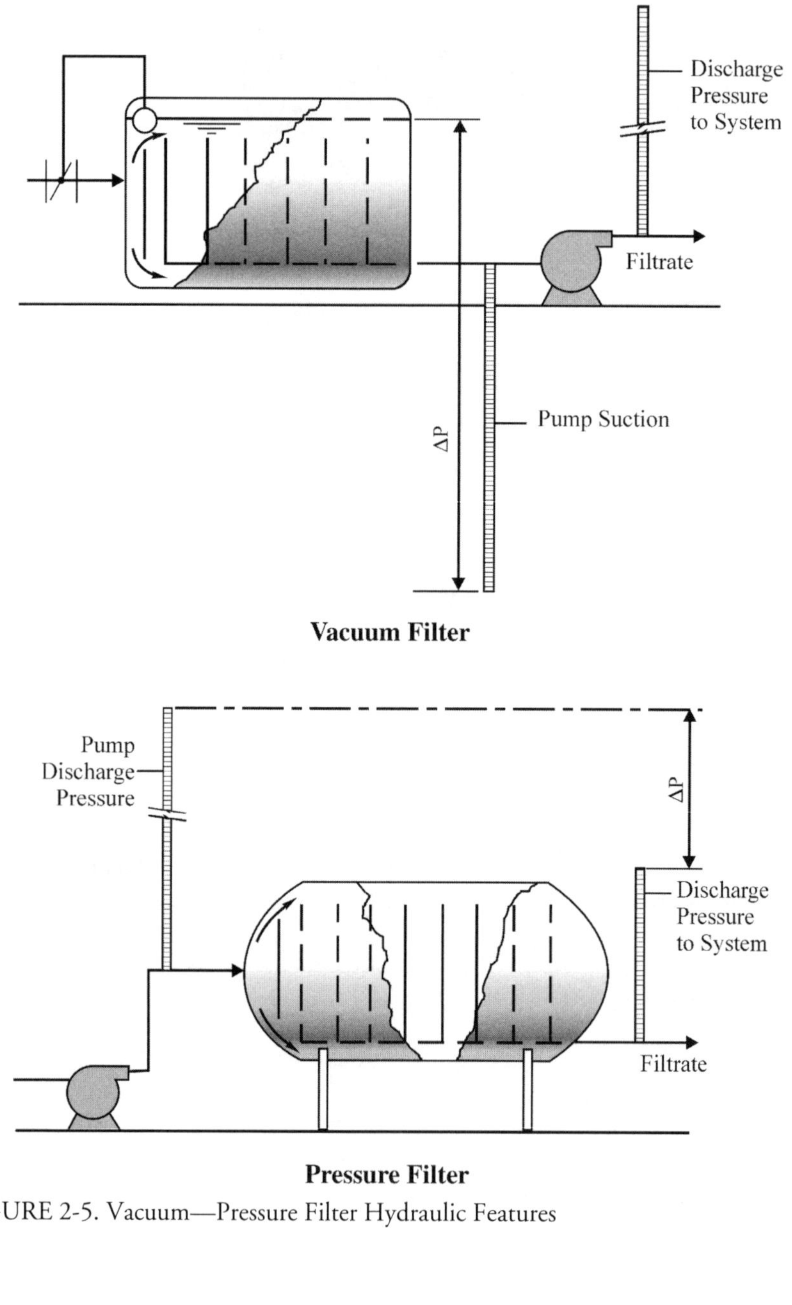

FIGURE 2-5. Vacuum—Pressure Filter Hydraulic Features

Pressure Filters

Operating under pressure, the principal advantages of this filter over the vacuum filter alternatives are related to the significantly higher differential head available. This higher head permits:

— Operation at higher filter application rates resulting in smaller, more compact filter units
— Longer filter runs, which reduce the use of precoat material and sluicing water because of less frequent cleaning cycles.

Theoretically, there would be almost no limit to the applied pressure and the filter rate adopted in operation. Since the filter leaf operates under a differential (external to internal), actual pressure on either side of the septum would have little influence on performance. Production rate and water quality would be the same whether a 20 psi differential was a result of a pump discharge of 25 psi and system inlet pressure of 5 psi (Figure 2-5) or an alternative of 50 to 30 psi across the filter. From a practical standpoint, however, there would be limits to both pressure applied and the differential provided. Higher pressures require heavier duty construction of the filter vessel and other components like hydraulic seals for moving parts penetrating the shell. Since power costs for pumping are one of the more significant features affecting the economic viability of DE filtration, this factor would limit the differential selected for operation. Based on designs and studies for small plants as well as for plants of over 100 mgd capacity, differential pressures of from 20 to 40 psi may be considered to be practical. Since most filtered water is provided disinfection in open gravity flow basins, pump discharge usually would be against a nominal back pressure. Referring to Figure 2-5, where source water is available from higher elevations, pumping costs could be reduced or even eliminated. The differential pressure selected for design, therefore, can be influenced by particular site conditions.

In comparison with vacuum filtration, the principal disadvantages of pressure filters would be:

— Higher capital costs for fabrication
— Higher maintenance costs for the tight seals required for mechanical and moving parts.

Since the pressure units are totally enclosed, general plant housekeeping costs would tend to be lower. Again, where filter capacity is not an issue, cost of construction and operation would be the major consideration in the selection of a pressure over the vacuum type filter for a particular installation.

Vacuum Filters

Open to the atmosphere, vacuum filters have the distinct advantages of permitting convenient observation of all cycles of DE filter operations and the opportunity to manually sluice cake from the leaves with use of a high-pressure hose stream. There are several disadvantages inherent in vacuum filter operations. Theoretically, total atmospheric pressure of around 34 feet would be available to generate flow if a pump suction could produce a complete vacuum. Based on the net positive suction head characteristics of pumps used for this service, 20 feet or around 9 psi would be a more practical limit to assume for the differential head on Figure 2-5. Even with the gravity flow alternative shown in Figure 2-5 for vacuum filter operation, pipeline friction losses might reduce the differential head available to from 20 to 25 feet. Compared to the pressure filter alternative, this available pressure limits vacuum filter capability as follows:

— Filter application rates would be limited to a maximum of about 1.5 gpm/sq. ft. of filter leaf. Application rates as low as 0.5 gpm/sq. ft. might be used to extend filter runs for higher particulate removal situations. Most vacuum filter installations are operated at rates of 1.0 gpm/sq. ft. or less.
— Lengths of run would be shortened, resulting in increased costs for the more frequent use of precoat and sluicing water.

In addition to the aforementioned limitations, the potential for air binding exists in vacuum filter operation. The vacuum suction can produce increased negative pressure in the filter cake, which is exacerbated by the increased flow velocity through the small interstices in the media. Under certain conditions of high dissolved air and/or gas concentrations along with low water temperatures, this negative pressure can cause gas to come out of solution. With the relatively thin layer of media or cake developed on the septa, the gas bubbles tend to "pop" the media bridging the septum openings. This action can disrupt the integrity of the cake to intercept particulate matter. At the least, the "popping" will release precoat material for an increase in particles and turbidity in the filtrate. Vacuum filters can be most vulnerable to this air/gas release during low water temperatures in winter. At many existing locations, however, low water temperatures, dissolved air, and media porosity conditions do not produce release of air. Where air release is a problem, pressure filtration would be the preferred mode to consider.

In addition to the benefits provided by the open top construction, vacuum filters have two other principal advantages over the pressure alternative:

— Capital costs for filter units can be significantly lower because the low pressures permit lighter weight construction and few mechanical parts and hydraulic seals are required.
— Simplicity of operation and maintenance permit use of less skilled plant personnel, and results in lower overall labor costs.

In many instances, the lower charges for capital improvement and the lower operation and maintenance costs can more than compensate for added precoat material and sluicing water costs. As mentioned before, properly operated vacuum type filters should produce filtered water quality equal to that of the pressure alternatives.

Containments

There are two basic requirements for any type of DE filter containment:

— Construction materials must be capable of resisting corrosion
— The filter structure must withstand the pressures and forces imposed by the hydraulic conditions and by the arrangements for mounting the filter body and the various attachments to the body.

Pressure Vessels

Carbon steel was used initially for pressure filters, and with proper maintenance, has provided reliable service. Recently, the predominant material for fabricating pressure filter vessels has been stainless steel. Many DE filters must treat somewhat corrosive surface waters. Moreover, filter screens can become clogged over time due to the deposits of iron, manganese, and/or organic matter, which require acid cleaning. These factors, in addition to length of service anticipated for water treatment filters and the relative reductions in stainless steel fabrication costs in recent years, makes stainless steel the preferred type of construction. Pressure vessels should be designed and fabricated in compliance with the ASME Boiler and Pressure Vessel Code—Unfired Vessel Sect. VIII, in addition to any other pertinent universal and local codes to ensure good workmanship and adequate strength.

Vacuum Filter Containments

Open to the atmosphere, vacuum filter assemblies may be installed in fiberglass-reinforced plastic, metal filter tanks, or in field fabricated concrete basins. Choice of containment will depend on the particular installation requirements and comparative costs. In general, concrete basins would be used for larger capacity, multiple DE filter applications where common wall construction is possible. Fiberglass reinforced plastic or metal tank construction may be used for both large and small filter units. All three types of construction require careful structural design to provide for the hydraulic and equipment loadings.

There are no specific codes for plastic or metal tank design and fabrication of vacuum filters. In general, acceptable structural analysis procedures for rectangular tanks should be employed. The analysis should determine the maximum stress and deflection that would occur for tanks of particular length to width ratios and height that consider the different combinations of wall thickness both with and without wall stiffeners. Since filter leaf supports are often attached directly to the containment vessel walls, it is best to keep wall deflection to a minimum. Depending on material of construction (plastic or metal), either vertical or horizontal stiffeners may be integrated into the design. All tanks would have angle type (or other structural member) stiffeners around the top edge. Most often metal fabrication would use standard angle or channel sections for both top and vertical stiffeners. Plastic fabrication would involve integrating the stiffening into the wall construction as shown in Figure 2-6. As in the case of pressure vessels, metal tank construction should use stainless steel.

An economic comparison should determine the choice between concrete, plastic, and metal construction. Most new vacuum-type filters today are fabricated of fiberglass-reinforced plastic.

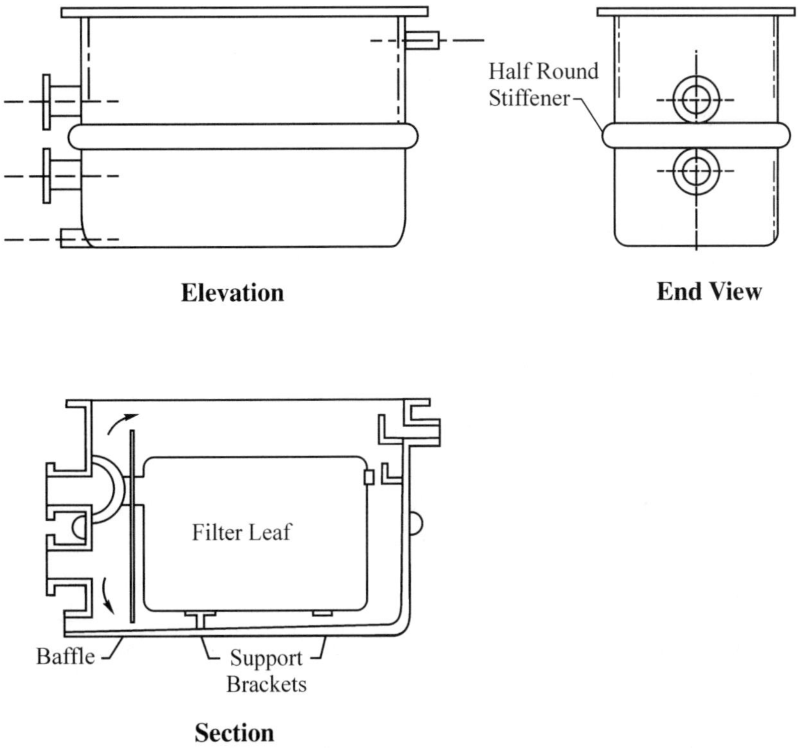

FIGURE 2-6. Typical Details—Vacuum Filter, Fiberglass Reinforced Polyester Construction

Filter Leaf Assembly

The filter leaf assembly must be the primary element of any DE filter. Particulate separation occurs on the precoat and cake held to the surface of the filter leaves by the flow.

As specified in the AWWA M30 manual, the following basic criteria must be observed in the design of filter leaves:

— Structural strength to provide a firm support for the filter-media cake.
— Adequate drainage area inside the element so the filtered water can easily exit the element.
— Proper construction of the septum to provide clear openings of proper size so the filter media forms strong, stable "bridges" over the

openings. The material of the septum should be capable of maintaining the integrity of the weave pattern to prevent distortion of opening size or shape with continued use.

— Corrosion-resistant construction materials for long life.

It is essential that in the assembly of multiple leaves, the individual leaves or the integral parts of individual leaves do not move or warp under increasing differential pressure. Such movement could result in the possible cracking of cake and breakthrough.

Figure 2-7 is a cutaway of a typical flat filter leave. Simply stated, a septum is the partition or divide between two cavities. In the case of the filter leaf, the septum is the woven cloth that divides the applied inlet water side from the internal collection chamber for the filtrate flow. The septum material may either be a tightly woven standard steel wire mesh or a monofilament polypropylene weave.

As shown in Figure 2-7, a heavy wire mesh provides the drainage space or chamber for collection of the filtrate flow. The internal chamber volume and configurations must not impede the flow to the outlet nozzle. The chamber material and the outside frame must provide sufficient

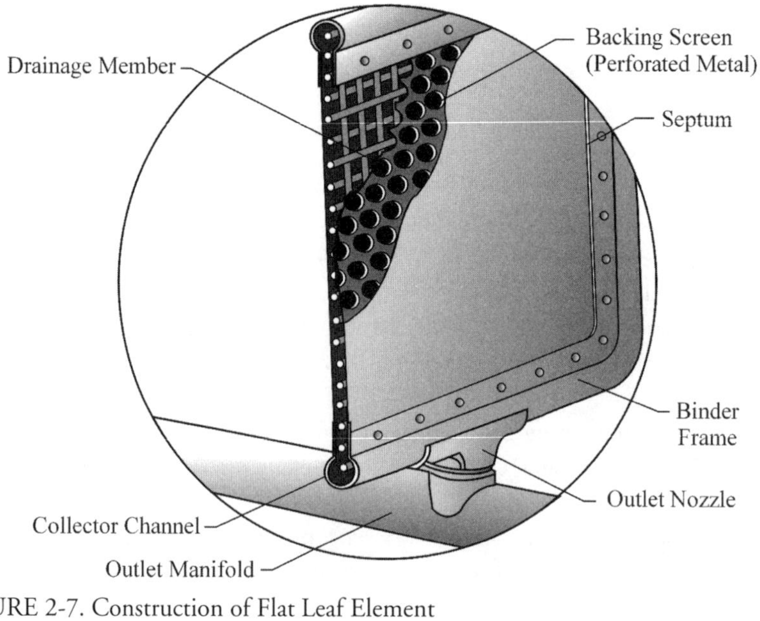

FIGURE 2-7. Construction of Flat Leaf Element

rigidity and strength to result in a reasonably flat frame that will not warp under the stress of differential pressure. A backing or intermediate screen is sometimes provided on either side of the central chamber or drainage member that will span the somewhat large interstices to prevent flexing of the septums. This stiffening support also prevents damage to the septum as it might flex under the application and release of pressure in operation. The intermediate screen may be a perforated metal sheet or a wire weave with a smaller wire spacing than the chamber material. The outlet nozzle must be large enough to handle the required flow without undue pressure loss. The outlet nozzle connecting to the outlet manifold must also provide a tight seat that permits easy removal of individual leaves.

The frame surrounding the leaf binds the entire assembly together into a unit. The frame provides a closure for preventing leakage or short-circuiting of the applied flow around the edge of the leaf, thereby bypassing the precoat. In addition, the frame is the principal structural member that provides rigidity for maintaining the shape of the leaf, for retaining the septum, and for mounting the outlet nozzle.

The design details and the type of materials used in the construction of filter leaves will depend on the type of filter and the method of sluicing the cake. In general, plastic materials are most often used in fabricating vacuum filter leaves; stainless steel for pressure type units. Some plastic materials also may be used in the fabrication of certain parts of pressure filter leaves. Flat filter leaves may be connected in a fixed position to the outlet manifold in both vacuum and pressure filter designs. Another option is available only for pressure filters where circular leaves are fitted with a central hub that fits around a center outlet manifold. In this option, the center manifold, which serves as the principal support shaft for the multi-leaf assembly, also rotates to permit sluicing by means of fixed water jets.

Drainage Member

Some of the principal types of drainage members available are shown in Figure 2-8. The first is expanded sheet metal, which is initially prepared with rows of slits on uniform centers. Assuming the slitted plate is in the horizontal position, material between the slits is alternately dimpled up and down to form loops of metal on both sides of the sheet. The particular advantage of this type of construction is that different size tubes or loops can provide the structural strength or rigidity required using thick-

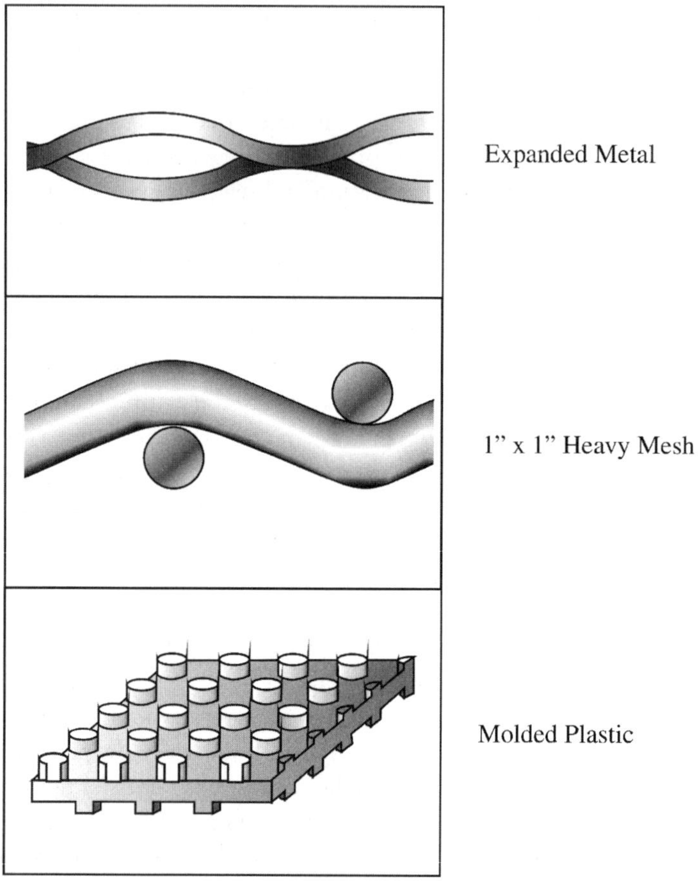

FIGURE 2-8. Drainage Chamber Options

nesses of stainless steel sheet between 16 and 20 gauge for both large and small filter leaves. Another common alternative is the heavy wire mesh using wires usually spaced one inch apart in both directions. Of the two, the wire mesh is the least expensive, but it frequently requires the addition of an intermediate or backing screen between the drainage chamber and surface septum for rigidity to reduce flexing of the septum. This addition can reduce the cost benefit of the wire mesh chamber, but the greatly reduced septa flexure can be an advantage.

Other types of drainage members are available. A chamber for a filter leaf may be a single molded sheet made from polypropylene. The drainage ways are provided by corrugations or other raised patterns in the molding.

Plastic drainage chambers are more common for vacuum filter leaves where differential pressure stress on the leaves is somewhat low. Plastic has sometimes been used for pressure filter applications where the size and span of the leaf may permit.

Other special chambers may be fabricated using spaced steel bars with round or wedge-shaped wires welded to both sides. The central bars provide superior drainage space and the outside wires may be spaced close enough to eliminate the need for intermediate screens.

Intermediate Spacing Screens

The basic purpose of the intermediate or backing screen is to span the too-wide spacing of drainage chamber patterns to limit the flexure of the outside leaf septa as differential pressure increases during filtration. Intermediate spacing screens may either be perforated metal sheets or wire mesh. Commonly, an 8 mesh (8 wires per inch) material is used for this alternative.

Leaf Septa

The septum is the most critical part of DE filter design and construction. For all practical purposes, the septum is the filter. The septum retains the precoat, which must bridge its openings. Special considerations in the selection of septa material include:

- — Cloth septa must be uniformly woven to permit an even deposition and bridging of the precoat.
- — The pattern of septa weave must allow the accumulated cake to drop freely during sluicing and must resist plugging and damage to maintain the integrity of the precoat support.

Many types of filter septa weaves are available. The most common metal weave, used for any type DE filter leaf, is the so-called 24 x 110 plain dutch weave. This wire cloth has high resistance to heavy sluicing pressures. This cloth may be calendared where the cloth passes through multiple compression rollers to produce a flat, smooth face. This helps to improve cake retention and release as well as wear characteristics.

Plastic cloth material used principally for vacuum filter leaves is available in a variety of weaves using polyester or polypropylene monofilament.

Filter Frames or Binders

The filter frame surrounds the leaf and binds all the aforementioned components of the leaf tightly together in an integrated assembly. The closure must be tight enough to be leak-proof using various methods that include bolting, riveting, and welding. Closures vary in cost and ease of repair (ease of disassembly) characteristics.

As discussed before, the filter frame may also provide the following in addition to its closure features:

— Rigidity to resist warping under pressure
— With tubular construction, an outside drainage way to assist conveyance of the collected flow to the leaf outlet
— A means of connecting the discharge nozzle connection to the collection manifold.

Some of the types of frame closures used in the fabrication of drinking water filter leaves are shown in Figure 2-9. The keyhole and "U" channel patterns are perhaps the most frequently used. Heavy duty tubular and

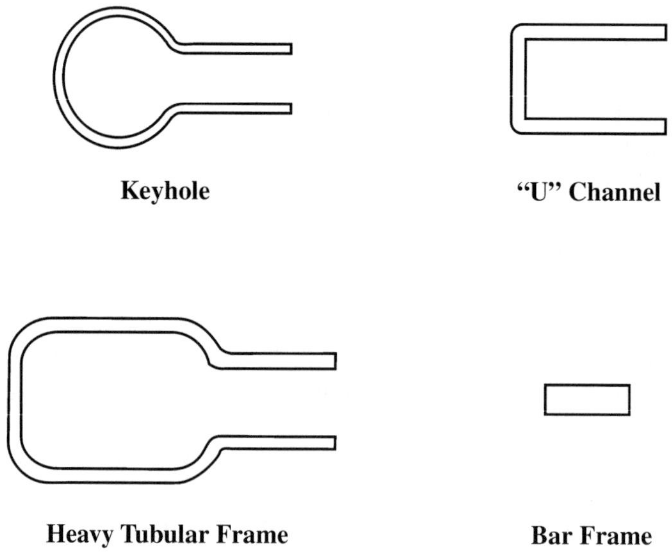

FIGURE 2-9. Binder Frame Options

bar construction would be used for larger-sized leaves where the structural stiffness provided by the frame is more critical.

Outlet Nozzles

The principal types of outlet nozzles used in filter leaf construction are shown in Figure 2-10. All are available in stainless steel for welding to the filter leaf frame. The most common outlet seal for fixed leaves is the "O" ring gasket type. This allows for convenient removal of an individual leaf for repair and the insertion of a temporary "O" ring plug if desired. The center hub shown would be used for a rotating leaf filter. The "O" ring gasket type and the center drain are also available in molded plastic. Plastic "O" ring nozzles are used predominantly for vacuum filter leaves either as separate units or integrated into the molding of the plastic drainage chamber.

Typical Leaf Assemblies

Almost unlimited combinations of the aforementioned components may be used in the assembly of DE filter leaves. Smaller-size screens usually are three layers with a central drainage member covered on both sides by the septum mesh. Larger screens may require additional stiffening for which a backing screen is used to develop a five-layer assembly. The backing screen will also help to limit flexure of the septum, which can be caused by a more open type drainage member. For the most part, vacuum filter screens are plastic cloth septa over a molded plastic drainage member of the "button" type shown in Figure 2-8 or ridged type options where the outlet nozzle may be molded as an integral part of the drainage member.

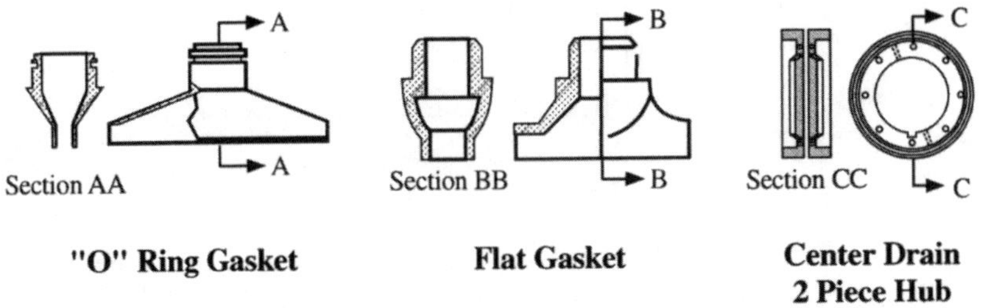

FIGURE 2-10. Outlet Connections

Baffled Inlets

The basic purpose of the baffled inlet is to introduce flow into the filter vessel in such a manner that the flow velocities do not remove or scour the precoat and accumulated cake from the filter leaf. The configuration of the baffles will depend on the type and size of filter vessel as well as the magnitude of flow. The baffle must provide a barrier to both dissipate the energy of the inlet connection velocity and to distribute that flow in a manner so that it is not directed at the surfaces of the filter leaves. Some of the various baffle arrangements that may be considered are as follows:

— *Vertical Tank/Flat Leaves.* Figure 2-11 shows a vertical baffle that can simply be a vertical section of pipe connected through the shell to the outside inlet. In this arrangement, the inlet velocity energy is dissipated in the vertical section and the flow is split to enter both the top and bottom of the vessel, above and below the filter leaves.

— *Horizontal Tank/Flat Leaves.* Figure 2-12 illustrates two baffled inlet alternatives for horizontal tank filters. In the upper Figure 2-12A, flow entering the end of a pressure filter vessel is directed against a flat

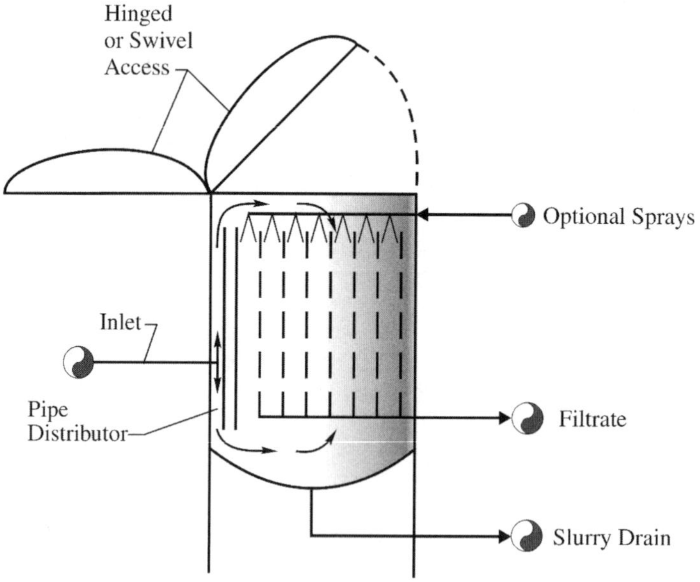

FIGURE 2-11. Vertical Fixed Leaf Filters

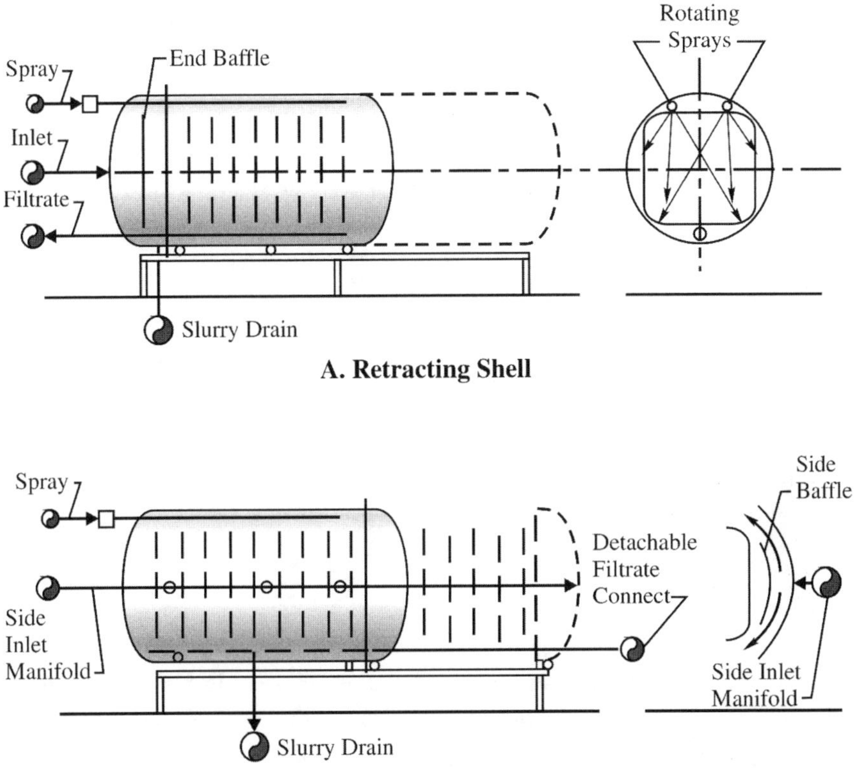

A. Retracting Shell

B. Retracting Leaf Bundle

FIGURE 2-12. Horizontal—Fixed Leaf Filters Arrangement

plate baffle. Flow energy is dissipated and is distributed around the edges of the baffle, sides as well as top and bottom. In this arrangement the baffle plate must be larger than the filter leaf. Not shown, this method may also be used for vacuum filters open to the atmosphere. When a pressure filter vessel is extended to accommodate a large number of leaves or when the vessel configuration does not permit an end inlet, another arrangement may have to be adopted. Here, the baffle plate may be shaped parallel to the outside circular shell, and flow is distributed into the vessel through the annular space provided (lower Figure 2-12B). Multiple connections to the shell are usually provided from an inlet manifold to improve longitudinal distribution in the longer vessels, which is illustrated for both the

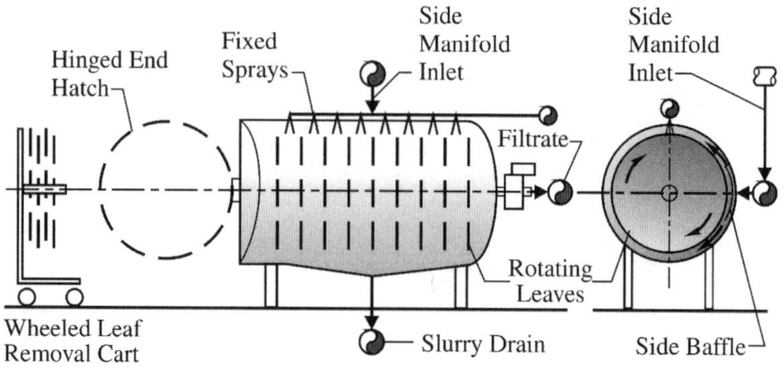

FIGURE 2-13. Horizontal—Rotating Leaf Filter

fixed leaf (Figure 2-12B) and the rotating filter (Figure 2-13) pressure filter configurations.

Outlet Manifolds

The most common position for the outlet manifold in DE filters is in the bottom of the vessel. Figures 2-11 and 2-12 illustrate this position for vertically mounted flat leaves in both vertical and horizontal filter vessels. Other manifold arrangements may be considered for different type DE filters. In vacuum filters open to the atmosphere, manifolds may be positioned on the side of the filter vessel with filter leaves connected at the top for easy access. Other pressure filters may employ a rotating leaf for improving the sluicing of the filter cake. Here the manifold is the rotating shaft and the leaves slide over the shaft for center discharge (Figure 2-13).

The basic requirement for all outlet manifolds is that sufficient flow capacity be provided. As a general rule, the velocity in the outlet manifold should not exceed 10 fps.

Sluicing Sprays

Sluicing spray designs are determined by the type of leaf assembly in the filters. Vertical filters have fixed sprays. Where the leaves are fixed in horizontal tank filters, spray nozzles are mounted on an oscillating pipe shaft to direct the sprays obliquely against the face of the filter leaves (Figure

2-12). The design should provide for high-pressure sprays between leaves to cover the complete filter leaf areas. A common problem with oscillating sprays is eventual clogging of the septum in areas where insufficient spray pressure is provided. Depending on the size of the filter leaf, two, three, or four nozzles may be installed to provide the necessary sluicing spray patterns and coverage between leaves.

For a rotating leaf filter, fixed sprays are attached to an overhead sluice water manifold to direct sprays in a fixed pattern radially across the circular filter leaf. As the center shaft rotates, all areas of the filter leaf are subjected to a high-pressure spray. The sluice spray system for the rotating leaf filter provides for more uniform cleaning in terms of pressure and coverage than is possible for the oscillating spray, fixed leaf design.

For smaller capacity filters like the open top vacuum filter or the vertical tank pressure filter (Figure 2-11), filter leaves may be sluiced manually using a high-pressure hose. The vacuum filter may be completely open or have a hinged cover. The vertical pressure filter is usually fitted with a quick opening hatch to facilitate optional manual cleaning. Manual cleaning is probably the preferred method for these designs, where spray from oscillating manifolds cannot be contained or where the spray header would interfere with access to the relatively small volume filters. An additional advantage of this method of sluicing is that the operator can observe all parts of the filter leaves, making it easy to see when the filter leaf is completely clean.

Pressure Filter Options

General

Pressure filter options may be defined by the mounting position of the tank, the method of mounting filter leaves, and the method of access to the filter leaves or the assembled filter leaf bundle for maintenance and replacement. Figures 2-11 and 2-12 show the more common arrangements for these options. The filters shown utilize flat leaves; all would be of stainless steel construction for the tank shell and internals. The vertical leaves connect to a central bottom manifold within the filter vessel with single push-on hubs at the bottom of each leaf. Leaves may be round, rectangular, or any version of a polygon required to accommodate filter internals such as spray piping and inlet baffles.

For the rotating leaf type (Figure 2-13), circular leaves are stacked in a common horizontal manifold. The manifold passes through the central hub of each leaf. The process path of flow is similar to that of the fixed leaf collected by the center manifold. An exterior rotating mechanism turns the leaf assembly for sluicing by fixed sprays mounted over the top of the leaves.

The following discussions for each of these four principal pressure filter options summarize the principal features. The advantages and disadvantages for each arrangement are listed for different application situations.

Vertical Tank—Top Access

Because of its vertical mounting, the vertical tank filter (Figure 2-11) requires less floor space than the other types of pressure alternatives. The unit's rectangular leaves are mounted vertically, connected by an "O" ring spigot into hubs in the central manifold at the bottom of the tank. Depending on the size of the unit, either a curved plate baffle or a simple pipe open at top and bottom may provide for balanced distribution of the inlet flow without disturbing the cake. For leaf inspection or maintenance, the top cover may be hinged for opening or may be provided with a handwheel lift cover for pivoting out of the way. Because of the vertical tank, the width of the individual rectangular leaves will decrease from the center of the shell outward. Vertical tank–vertical leaf filters are available with total leaf areas varying from around 20 sq. ft. to almost 1,000 sq. ft., but are used more in the smaller area–lower capacity range.

Vertical filters may be fitted with fixed sprays between the leaves to sluice the accumulated cake. Smaller diameter units may be manually cleaned with a high pressure hose. Because of the height of the unit, individual units or batteries of units must be provided with raised platforms, preferably 3 to 4 feet below the top of the open tank for convenient access.

ADVANTAGES:

— Requires the least amount of floor space
— Provides easy access through top
— Smaller units may be manually sluiced
— Less mechanical features and four-legged mounting makes unit less costly than alternative pressure types.

DISADVANTAGES:

— Multiple widths of vertical leaves requires maintaining several sizes in inventory.

— With access from the top, filter leaf positioning into a bottom socket may be somewhat difficult.

Horizontal Tank—Retracting Shell

With this arrangement (Figure 2-12A), all connections may be made through a fixed head including inlet and outlet flow, sluiced slurry drainage, and the oscillating spray water supply. As the length of tank increases, the single entry baffle in the fixed section may not be sufficient for uniform flow distribution or may require larger-than-desired velocities to move flow front to back. In this case, multiple inlets would be made to the movable shell with a quick disconnect to discharge against side baffles. To gain access to the leaves, a holding gland is first rotated to break the lock and seal between the fixed section. The shell is then retracted with rollers on a track. Riding on the shell, rollers on the end of the leaf bundle hold the leaf assembly horizontal as the shell is retracted.

Retracting shell units are available with total leaf areas of up to 2,000 sq. ft. This type unit is probably best adapted for leaf areas of up to 1,000 sq. ft.

ADVANTAGES:

— All process connections are fixed to one head; no disconnects are required to open the vessel (except for side baffle inlets).

— All leaves can be inspected or removed individually from the filter.

DISADVANTAGES:

— Additional floor space is required to permit retraction of shell.

— For a given leaf size, a larger diameter shell is required to fit the collection manifold below the leaves.

— Leaves are sluiced by rotating spray headers, which require moving parts, bearings, and bearing seals subject to wear. Spray intensity over the leaf area varies, which may affect uniformity of cleaning.

— Discharge from a leaf occurs from a single point along the bottom edge. For large leaves, uniform flow distribution within the leaf may be inhibited as water must travel longer distances to reach the discharge point.

— For large leaves, the capacity of a single leaf is limited by the size of the discharge point. Increasing the size of the discharge connection beyond 2 inches would increase the leaf spacing and hence the shell length of a filter.

Horizontal Tank—Retracting Bundle

This unit (Figure 2-12B) is similar to the aforementioned horizontal option except that the tube bundle rather than the shell is retracted. With this arrangement, all inlet, spray, and sluiced slurry connections are made to a fixed shell and remain intact. Only the outlet manifold connection must be provided with a quick disconnect. This arrangement is particularly well-suited for extended shell lengths to provide large leaf areas.

To retract the bundles, a holding gland is first rotated to break the lock and seal with the fixed shell. The bundle moves on rollers riding on two tracks with the end of the bundle supported by another set of rollers.

This type of unit is available in sizes of more than 2,000 sq. ft. total leaf area. It is probably best adapted for ranges of more than 1,000 sq. ft. because extending the length does not influence operating effectiveness.

ADVANTAGES:

— Fixed shell permits multiple inlet points along the length of the filter to improve flow distribution.
— All leaves can be inspected or removed individually from the filter.

DISADVANTAGES:

— Additional floor space is required to permit retraction of leaf/manifold bundle.
— For a given leaf size, a larger diameter shell is required to fit the collection manifold below the leaves.
— Quick disconnect filter outlet couplings are required to permit the leaf/manifold bundle to be retracted.
— Discharge from a leaf occurs from a single point along the bottom edge. For large leaves, uniform flow distribution within the leaf may be inhibited as water must travel longer distances to reach the discharge point.
— For large leaves, the capacity of a single leaf is limited by the size of the discharge point. Increasing the size of the discharge connection beyond 2 inches would increase the leaf spacing and hence the shell length of a filter.

— Leaves are sluiced by rotating spray headers, which require moving parts, bearings, and bearing seals subject to wear. Spray intensity over the leaf area varies, which may affect uniformity of cleaning.

Horizontal Tank—Rotating Leaf

As shown in Figure 2-13, the inlet, spray, and sluiced slurry drain connections are made to the fixed shell of the rotating filter. Outlet flow, however, is through the rotating manifold fitted with a mechanical seal. Leaves have a central hub for sliding on to the central manifold. These hubs have flat machined-bearing surfaces that seal the leaf assembly as the stack is held together with an end compression nut. Single fixed sprays are mounted at the top between leaves for sluicing the cake as the assembly is rotated.

Leaves must be extracted individually either by hand or by extension hooks to grab loop brackets mounted on each leaf.

The extracted leaves may be removed onto a mandrel mounted on a forklift or other type of wheeled truck to protect them from damage.

ADVANTAGES:

— Fixed shell permits multiple inlet points along the length of the filter to improve flow distribution.

— All process connections are fixed to one fixed shell; no piping disconnects are required to open the vessel.

— Center collection manifold and round leaves permit the smallest diameter shell for a given leaf size.

— Less floor space is required, as there are no retracting parts.

— Leaves are sluiced by a single fixed spray header. No moving parts are associated with the header. Coverage of leaf is consistent.

— Symmetry of leaf area about the collection header provides even flow distribution to the center manifold regardless of leaf size.

— Center discharge permits multiple leaf discharge points around the circumference of the common manifold so leaf spacing is not affected by leaf size.

DISADVANTAGES:

— Leaves cannot be inspected or removed from the filter individually. Inspection of all leaves would require removal of all leaves.

— Handling of leaves requires an operator entering vessel or the use of special pull hooks.

— Central collection manifold requires support during operation and temporary support during leaf inspection and maintenance.
— Leaves can be damaged if rotated while vessel bottom contains sluiced solids.

Vacuum Filter Options

Unlike the pressure filter, most vacuum filter configurations are similar. The principal differences exist in the material used for construction of the holding basins and for fabrication of the filter leaves, the position of the collection manifold for the filter leaves and the option of manual or oscillating spray sluicing. In general appearance, most vacuuming filters look alike.

As mentioned previously, vacuum type filter equipment may be installed in either concrete or reinforced plastic basins. Typical fabrication details for fiberglass reinforced polyester fabrication shown in Figure 2-6 indicate two bands of horizontal stiffeners for structural support of the sidewalls. Stiffening is essential to limit deflection of the sidewall to cause movement that could affect the integrity of the cake accumulated on the filter leaves. A 48-by-60-inch filter leaf is probably the practical maximum for vacuum filter installations. In the design shown, the dimension "w" may be increased to accommodate from 4 to 12 filter leaf elements for filter areas of from 160 to 480 sq. ft. At the conventional maximum application rate of 1.0 gpm/sq. ft., these filters can treat from 160 gpm to 480 gpm or from 230,000 to almost 700,000 gpd. For the open-top vacuum filter, an overhead monorail and hoist may be useful for removal of larger-sized screens. Smaller leaf configurations are available from several manufacturers to adapt basin size or capacity to particular situations.

Figure 2-14 shows the two options for mounting filter leaf assemblies. The bottom outlet is similar to that for pressure filters. A central bottom spigot on each leaf fits into a socket in the manifold sealed by flat gaskets or "O" ring gaskets. The alternative vertical edge or top-side manifold mounting has the advantage of placing the connection in a location that is visible and accessible at all times. As shown, both methods require top-edge bracing to maintain spacing and prevent movement. The edge outlet arrangement also has a central bottom support for each leaf.

The only other significant option available for vacuum filter arrangement is the method of sluicing the cake. All vacuum filters would be posi-

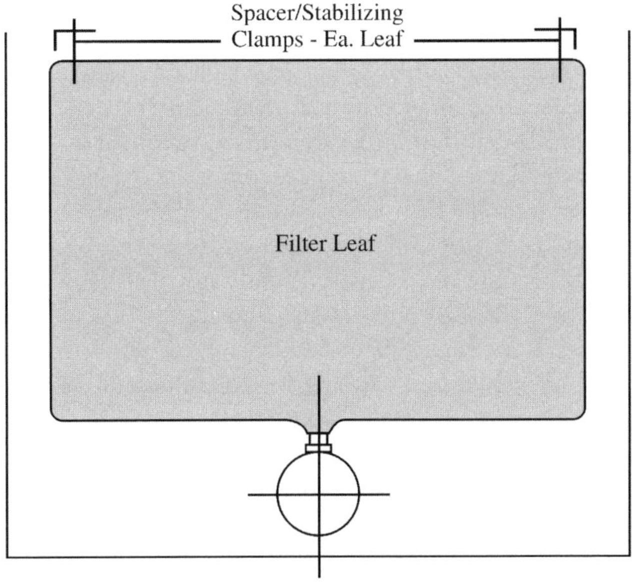

Bottom Manifold

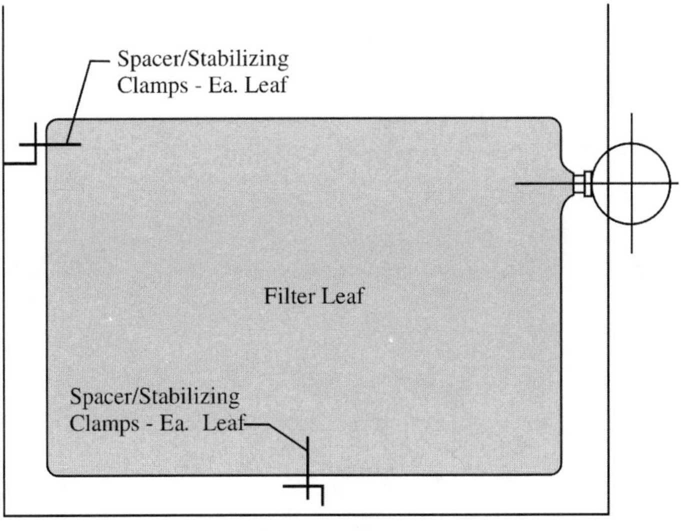

Side Manifold

FIGURE 2-14. Optional Vacuum Filter Leaf Assemblies

tioned with access to the side where the leaves are connected or braced. An operator, therefore, can clearly see between the leaves of a drained filter unit. Cake can be sluiced manually from the leaves using a high-pressure hose nozzle. Because most vacuum filter installations are low in capacity, the manual method of cleaning predominates. In some ways, it is the ideal way to clean the filter. The operator continues the spray to ensure that all cake is flushed from the filter leaves and flushed down the drain.

The optional method of cleaning would be with oscillating sprays similar to the pressure filter. A permanently installed spray header would be installed over the top center of the filter leaves. These sprays usually require a cover over each unit to contain the rebound. In any case, the oscillating spray would be used only where multiple units and a limited operating staff would require this labor-saver.

3 Principal Performance Features

Design of a diatomaceous earth filter train should be based on actual performance features determined in a pilot or other research program. Basically, the primary function of a DE filter or any porous filter is the removal of particulate matter, which has been related to the reduction of pathogens in drinking water. The trends in the federal SDWA regulations to lower levels of particulate matter and to broaden considerations of disinfection byproducts now demand comprehensive investigation of potential treatment requirements. While research is almost mandatory today, the actual performance of DE filters in drinking water treatment operation or research elsewhere would help in understanding some of the typical problems that must be faced in the testing and in the design of full-scale facilities. Fortunately, a wealth of performance data is available from extensive studies conducted in City of New York drinking water treatment investigations and in other programs, some of which are reviewed later.

Principal Particulate Features

Measure

For decades, turbidity has been used as the measure of the particulate or suspended solids content of drinking water. At best, turbidity may be considered to be a surrogate measure of solids content. Particle counters that measure the actual numbers of particle in a range of sizes above 2 microns are providing other, more realistic means of determining filter particulate removal efficiency. In fact, studies indicate that there is no direct relationship between turbidity and particle count. Particle counts for product waters from DE and granular filters, for instance, are quite different even for identical turbidity levels.

Because the turbidimeter is relatively simple in concept and is easier to maintain, it should continue to be used for many years before the more costly, complex particle counter can be adopted as the universal measure. DE filter solids removal performance, therefore, is discussed in terms of both turbidity and particle count.

Size Relationships

Perhaps the most significant feature of solids removal in DE filtration is the relationship of the size of source water particles to be removed compared to the size of the DE media particles. Table 3.1 shows the count of source water particles in a large northeastern U.S. water supply impoundment. Counts are shown for raw water, after ozonation and passage through a BAC media contactor. As is discussed later, in addition to raw water, ozonated and BAC contacted water may precede filtration, depending on overall treatment needs.

Particle counts are indicated in the table for four ranges of sizes and those over a maximum of 15 microns for the three possible sources compared to that for Hyflo, a common media used in DE filtration. It is evident that, while the DE media has more than double the amount of particles (by weight) than that in the source waters for sizes above 15 microns, there is a similarity in sizes. Other determinations for this same raw water source of supply have indicated that 99% of the particles are less than 36 microns in size most of the time. This may be compared to Hyflo where 83% of the particles (by weight) are less than 36 microns in size. These

TABLE 3.1. Comparison Particle Count and Weight by Size: Source Waters and DE Hyflow

		Size (microns)					
		2–3	3–5	5–8	8–15	>15	Total
Raw	Count	1,474	4,568	1,975	837	100	7,480
	% by Wt	0.8	10.0	18.5	43.4	27.3	
O_3	Count	1,251	3,209	1,132	503	100	6,195
	% by Wt	0.9	9.7	14.8	36.4	38	
BAC	Count	283	566	152	67	11	1,079
	% by Wt	1.5	13.2	15.4	37.5	32.3	
DE Hyflo	% by Wt	2.0	6.0	7.0	18.0	66.0	

data may be compared to the sizes of granular media used in rapid sand filters. Effective sizes of granular media may range anywhere from 0.45 mm (450 microns) to 1.40 mm (1400 microns).

Another somewhat significant feature of DE media is the range of particle sizes. A measure of an effective granular media for rapid sand filtration is the uniformity coefficient (UC), which is defined as the ratio of grain sizes (by weight) as follows:

$$\mathrm{UC} = \frac{60\% \text{ are finer (by wt.)}}{10\% \text{ are finer (by wt.)}}$$

A well-screened rapid sand media will have a limited range of particle sizes indicated by a uniformity coefficient of 1.5 or less (1.0 being virtually all equal particle sizes). Using the same measure, the uniformity coefficient of Hyflo DE media would be almost 4.0, indicating an extremely wide range of sizes. The nature of loose particles applied in slurry-like DE precoat and body feed is that the smaller particles will "nest" in the interstices formed by the larger shapes. This tends to reduce porosity or the size of flow path through the media.

The relative sizes of source water particles (Table 3.1) tend to exacerbate the "nesting" situation. In general terms, the mostly smaller sized source water solids will add to those smaller media particles that fill in the interstices formed in the precoat, which reduces porosity further.

The median pore size of Hyflo media has been determined to be about 7.0 microns; the predominant amount of the source water particles in Table 3.1 are greater than 7 microns in size. Unfortunately, little work has been devoted to determining pore sizes in granular media. The average pore opening in 1.0 mm (1,000 micron) granular media would be more than 100 microns. A negligible amount of the source water solids shown in the table are that magnitude in size. These size comparisons indicate why the principle solids separation mechanism in DE filtration is entrapment and in granular filtration, attachment. Separation of the much smaller colloidal-sized particles would be a problem common to both DE and granular filtration.

Source water particles in the size range of 2.0 to 15.0 microns are of particular concern in drinking water treatment. *Giardia* and *Cryptosporidium* cysts fall in this size range. The ability of a filter to remove particles in this range, therefore, would be a measure of its effectiveness in removing these pathogenic cysts. Removal by entrapment, inherent in DE filtration, would be more positive than separation by attachment in granular filters. A trapped particle cannot move; an attached particle, held by magnetic-like forces, may be dislodged by the impact of hydraulic flow.

Particulate Character

The character of particles affects removal as much as size. Source water particles may range from inert silt carried in river flow to those in impoundments or lakes that may be algae or inert solids covered with biologic growth. The two major concerns about removing solids in DE filtration are compressibility and the electrophoretic charge potential.

Particles that are compressible like algae, plant detritus, or slime-covered inerts will accumulate on the precoat in the separated solids cake. The hydraulic force generating flow and holding the cake to the leaf septum will also force the bound water out of the particles. This will cause the cake to compress, which can reduce the porosity of the overall cake significantly. The greater the proportion of compressible solids, the greater and more rapid the reduction in porosity or flow path area. On the other hand, dense silt, with particle sizes similar to that of the DE media, will not compress and the porosity of the cake would be maintained just as if all particles were diatomaceous earth.

Most NOM solids or solids coated with NOM have negative charges. The repellent forces generated between particles by the charges will inhibit solids separation in filtration by attachment. While most solids separation in DE filtration is by entrapment, colloidal matter must be removed by attachment to the media, similar to granular filtration. Colloidal color can be a significant problem in many source waters, which would require some kind of conditioning to reduce the effect of the negative charge. The options available for contending with negative-charged source water particles are discussed later.

Head Loss and Body Feed

Hydraulic head loss across an operating DE filter is increased between inlet and outlet connections. Total head loss is the sum of the following:

— *DE Filter Unit.* The loss in the inlet and outlet flow connections plus the internal losses, including those of the inlet baffle, the filter leaf, the filter leaf manifold, and the loss in the precoat.

— *Accumulating Cake.* Loss in the accumulating cake, which includes those generated by the raw water solids and body feed.

Total head loss through an operating DE filter is a combination of laminar and turbulent flow conditions. Flow through the clean precoat is laminar, where the increase in the unit head loss would be directly proportionate to an increase in flow rates. Flow through the physical elements of the filter is more turbulent in nature, a combination of losses through inlets, outlets, pipelines, and the various flow paths. Turbulent flow also exists in the accumulating cake where the separated solids significantly reduce the porosity of the combined separated solids/body feed cake. E.R. Bauman expressed the loss through a DE filter cake, at any point in time, as follows:

$$\begin{aligned} H &= h_1 \text{ (precoat)} + h_2 \text{ (body feed layer)} \\ &= K_3QW_1 + K_4C_DQ^2t \times (8.33 \div 10^6) \end{aligned}$$

Where:

- H = Total head loss across filter cake, ft. of water
- Q = Rate of filtration, gpm/sq. ft.
- W_1 = Weight of precoat, lb/sq. ft.
- K_3 = Permeability characteristic of the precoat filter-medium layer, min-ft^5/lb.gallon
- C_D = Concentration of diatomite in body feed, ppm
- t = Length of filter run, minutes
- K_4 = Permeability characteristic of the body feed plus suspended solids layer, min-ft^5/lb. gallon.

Compared to either the head loss in the filter internals or the accumulating cake, the precoat loss is relatively insignificant. Therefore, to determine the relative rate of head loss or total head loss through a DE filter for different application rates, it would be sufficient to say that head loss is proportional to Q^2.

Body Feed Influence

At a fixed flow rate, the head loss through the filter elements and through the precoat are fixed amounts. Head loss through the accumulating mix will vary according to the character of the raw water solids and the proportion of DE body feed added to the inlet flow.

The body feed provides a matrix for the accumulating filter cake as it surrounds the raw water matter. The filter media itself is relatively unyielding, which is effective in maintaining the porosity of the cake. As a result, with sufficient body feed the head loss should build up more as a function of cake thickness than as a result of decreasing porosity. Some reduction of porosity must result from the separated raw water solids, but such reduction is minimized significantly with judicious use of body feed.

Where the raw water solids tend to be organic in nature and compressible, as the run progresses the increasing head loss can increase the compression of the cake. Cake head loss for a pressure type DE filter in drinking water treatment may vary from 20 to 35 psi; that for a vacuum type filter, from 7 to 8 psi. At the start of the filter run, the fixed loss for both type filters would be in the order of 2 to 10 psi, depending on the filter configuration. Correct body feed, therefore, is especially important where compressible matter is removed in DE filtration to ensure efficient filter runs.

Figure 3-1 shows the effect on head loss from varying the amount of body feed to compensate for the raw water characteristics. Body feed varies from 0 to more than 27 mg/l. Two distinct features are illustrated in the Figure.

1. Length of run increases with increasing the body feed ratio to 20 mg/l. At 27.3 mg/l, the length of run decreases.
2. With no body feed, the head loss variation has a curved characteristic indicating changing turbulent flow conditions as increased head loss compresses the cake to progressively reduce pore size. As the body feed is increased, the head loss curve straightens out to reach a straight line at 20 mg/l, an indication that the only change in head loss is due to the increasing thickness of cake.

In essence, at a body feed of 20 mg/l, the impact of the separated raw water matter is eliminated sufficiently to be negligible. At any body feed rate above 20 mg/l, the DE porosity prevails, but at the higher than necessary feed rate, the flow path through the cake is increased more rapidly to shorten the run. The basic objective in establishing the most efficient body feed ratio, therefore, is to provide just enough body feed to eliminate the influence of compression of the separated source water solids.

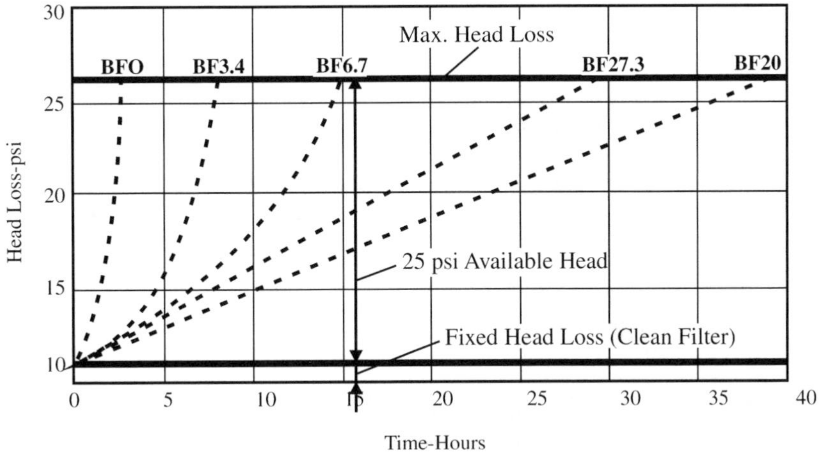

FIGURE 3-1. Effects of Varying Body Feed Addition

It is noted that the design of the physical elements of the DE filter should be such that the head loss for the anticipated range of application flow rates is minimized. Filter design would be the responsibility of the manufacturer, but the flow requirements should be defined in the specifications.

Figure 3-2 shows the head loss progression in a pilot-scale DE filter at a somewhat high flow application rate of 3.5 gpm/sq. ft. The ideal straight line progression is indicated before head loss reaches 25 psi after 51 hours of operation. In this installation, the flow control for the low 3.5 gpm rate is subject to "hunting" indicated by the somewhat erratic automatic plot. Such hunting would not exist to this extent for the flows in full-scale units. Despite the erratic nature, the straight line head loss progression is clearly evident.

Flow Rate Influence

Figure 3-3 shows the comparison of head loss curves for a DE filter operating at two different filter application rates under otherwise identical conditions. These data confirm the Bauman formula for accumulating cake head loss plus the loss through the filter elements. The rate of progressive head loss increase and the time to reach the total available head loss of 25 psi are a function of Q^2. At an application rate of 2.0 gpm/sq. ft., the length of run was four times that for a rate of 1.0 gpm/sq. ft. It is noted that the Figure shows only the progressing head loss, not that for

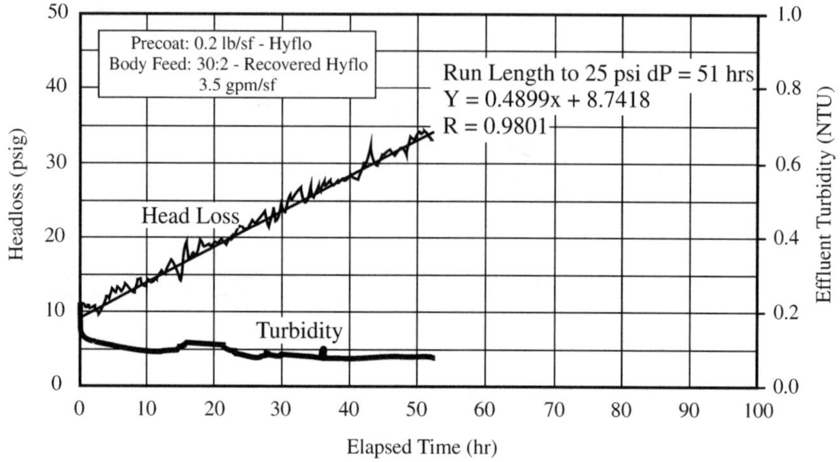

FIGURE 3-2. Headloss, Effluent Turbidity vs. Elapsed Time

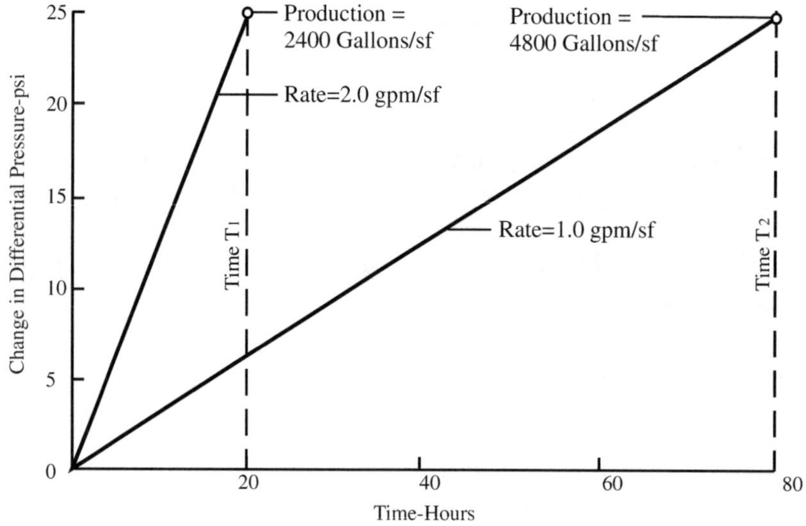

FIGURE 3-3. Influence of Application Rate on Production Runs

the filter elements and precoat loss. The data is from the operation of two parallel, full-scale filter units.

Enhanced Solids Removal

Enhanced solids removal in DE filtration simply involves compensating in treatment for the inhibiting characteristics of the raw water solids to be separated. In summary, the principal problem features of such solids include:

— Size ratio that tends to nest or fill in the DE precoat interstices
— Compressible solids, usually of biologic or organic origin
— Negatively charged particles of NOM, particularly those of colloidal size that must be removed by attachment in the cake pores.

The first two features may be corrected through proper body feed. Reducing the relative compressibility of organic matter, however, will reduce the amount of body feed required. Unless the negative charges of colloidal-sized particles are neutralized, the particles will simply pass through the cake. This is indicated by the limited ability of DE filtration to reduce colloidal color without treatment compensation.

Ozone oxidation of the inlet raw water has been proven to be effective in both reducing particle compressibility and particle charge. Coating the filter precoat with positively charged alum has also enabled trapping negatively charged particles. Typical performance details of these treatment measures are discussed later. Performance data is from DE filtration of a northeast U.S. impounded water supply. The overall reservoir system may be considered to be subjected to moderate eutrophication. Raw water solids are predominately organic in nature, which has been indicated by the distinct slime-like character of the DE cake where no preconditioning was employed. There is little to no silt in the reservoir outlet flow. Most of the inert solids are considered to be amorphous matter. Depending on the time of the year, the raw water solids may include slime-covered inert solids, leaf and plant detritus, and various types of algae.

Ozone Conditioning

Preozonation changes the character of the inlet solids by dissolving slime growth and breaking up certain biologic agglomerates. Similar to its disinfection mechanism, it literally destroys biologic matter. Once separated from its slime or biologic covering, inert colloid solids lose their negative charge. Preozonation, therefore, can have the multiple effect of reducing the content of compressible solids and neutralizing negatively charged particles.

Table 3.2 shows how the volume of solids is reduced through ozonation. In this particular research program, six ozone contact columns were provided to study the change in solids content. Particle counts were used to calculate suspended solids volume in each stage of contact. The Table shows inlet solids, maximum solids content in an intermediate stage, and the effluent solids.

Ozonation also has a micro-flocculation effect in particle conditioning. When the negative charge is removed, the particles agglomerate. In addition, where dissolved iron or other metals may exist, hydroxides can be formed to enhance the agglomeration action. As indicated in Table 3.2, the solids content in passage appears to first increase then decrease to the outlet. With both micro-flocculation and organics dissolving occurring at the same time, the increases in solids is more apparent at the lower 6 minutes' detention. Except for what is probably a slug of collected solids in the first run at 10 minutes' detention, the rate of the dissolving action does not permit significant solids buildup. For all runs shown, the effluent solids content is less than the inlet with little apparent impact from the rate of ozonation or detention time. The impact of different ozone rates on reducing compressible matter, however, is more apparent when the length of run or rates of head loss progression are compared in Figure 3-4.

Head loss progression for individual DE filter runs are shown on Figure 3-4 for no preozonation and for ozone rates of 0.5, 1.0, 1.5, and 3.0

TABLE 3.2. Particle Volumes Through Ozonation

Flow (gpm)	Detention (minutes)	Ozone Rate (mg/l)	Particle Volume (ppm)		
			Raw Water	Maximum/Col.	Last Col (F)
720	6	0.5	1.61	1.78	1.30
720	6	1.0	1.61	2.66	1.24
720	6	1.5	1.61	2.55	1.20
440	10	0.5	2.90	7.61	1.30
440	10	1.0	2.90	1.78	1.30
440	10	1.5	2.90	2.04	1.36

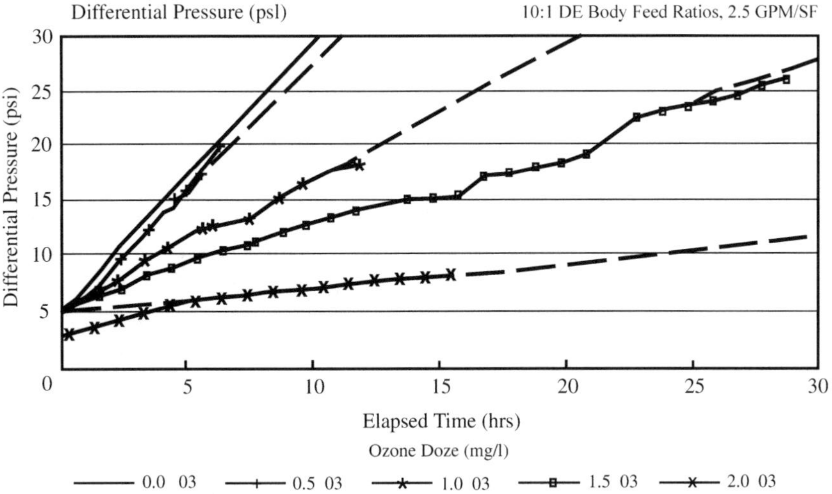

FIGURE 3-4. Loss of Head, Various Ozone Rate vs. Elapsed Time

mg/l. Length of run (projected in some instances) to 25 psi head loss is as follows:

Ozone Rate (mg/l)	Length of Run (hours)
0	10.0
0.5	11.0
1.0	20.5
1.5	35.0
2.0	111.0

While it is interesting to observe the potential impact of the 2.0 mg/l preozonation rate, DE filter runs of more than 100 hours would not be practicable. The potential cake buildup would probably be more than the filter leaf spacing would permit.

Except for potential breakthrough of colloid-sized particles, preozonation should not otherwise influence the product water turbidity or particle count.

Alum Coated Media

To call the procedure "alum coated DE media" is almost a misnomer because the media is actually coated with aluminum hydroxide. The medium is prepared by first mixing DE with alum before adding soda ash to the slurry. The hydroxide coats the media grains to provide a somewhat gelatinous characteristic as well as a positive charge.

The best way to track the effectiveness of alum coated media is with the removal of color or, to be more exact, the removal of colloidal color. When first suggested as an option to improve the removal of colloids in DE filtration, alum coated media was to be used both for precoat and body feed. In a somewhat intensive research program for treatment of the City of New York Croton supply, it was demonstrated that alum coated DE for body feed was not necessary. Use of coated body feed significantly reduced the length of filter run. Another feature demonstrated in this research was the benefit of a two-layer precoat. If coated media is used as precoat, the cake does not readily separate from the filter leaf in sluicing. This was corrected by first precoating with virgin DE at a rate of 0.05 lbs/sq. ft., followed by a precoat of 0.15 lbs/sq. ft. of alum coated media. The result was a precoat that readily separated in the sluicing cycle and more than enough coated media to remove the colloids. Uncoated virgin DE or recovered DE was used for body feed. Table 3.3 shows results of typical test runs. The coated DE was considered as an option to preozonation of the filter inlet, which had also been demonstrated to be effective.

As shown in Table 3.3, the alum coated DE in the precoat removed the apparent color. The added use of preozonation in Run 3 extended the run length by conditioning the compressible solids in the inlet flow, but was not needed for the removal of colloidal color. Removal of turbidity for the runs in the table and other tests showed no apparent improvement for the addition of preozonation. It was concluded that where color levels are moderate to low, the use of alum coated DE precoat will effectively remove most colloidal color and is just as effective in reducing turbidity as preozonation of the inlet flow.

TABLE 3.3. Filter Runs: Alum Coated Precoat

Ozone (mg/L)	Turbidity (NTO)		Color		Length Run (hours)
	In	Out	In	Out	
0	1.30	0.11	16	0	22.5
0	1.35	0.19	24	0	26.5
0.5	2.20	0.11	22	0	34.0

Notes: Filter application rate, 3.0 gpm/sq. ft., was identical for all runs. Two Stage Precoat: First Stage, 0.05 lb/sq. ft. Virgin DE; Second Stage, 0.15 lbs/sq. ft. alum coated.

Breakthrough Potential

Once a DE filter run has been initiated, there is practically no breakthrough of raw water solids as long as the flow that holds the cake to the leaf septum is not interrupted. Most of the low turbidity and particle count that is indicated in the effluent is probably fine DE material from the precoat that has not stabilized in the bridging of DE particles over the openings in the septum. Challenge tests using *Giardia* cysts in the 3- to 15-mm size range have demonstrated this advantage of DE filtration. Five to seven log removals of the cyst solids have been demonstrated in these tests.

Figure 3-5 shows a comparison of typical turbidity breakthrough patterns of DE and rapid sand filtration. There is elevated turbidity breakthrough at the start of both type filter runs, but higher for rapid sand filtration. As the filter runs stabilize, the solids or turbidity breakthrough should gradually decrease as the run progresses for each type filter run. The nature of rapid sand filtration is such that, if available head loss does not necessitate terminating the run, there will be breakthrough at some point in time. Separation is predominately by attachment to the filter

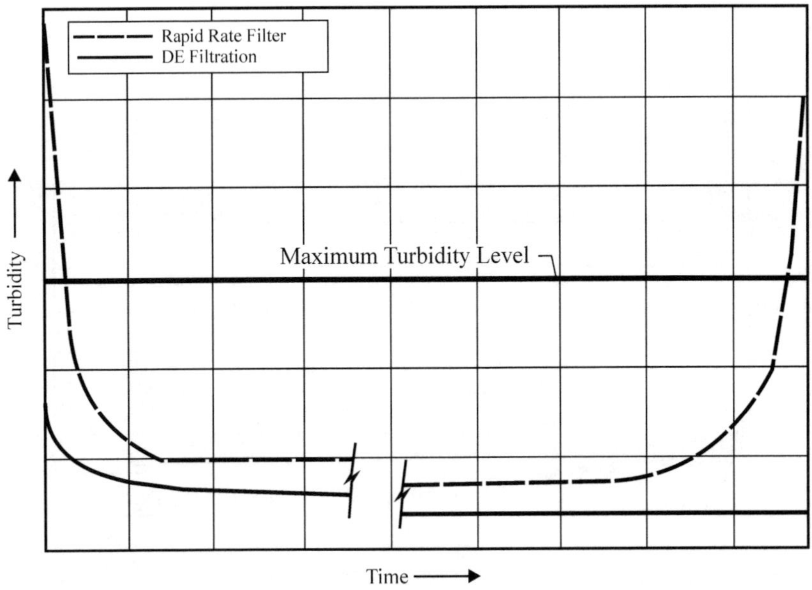

FIGURE 3-5. Turbidity Variations vs. Time

media and the buildup of separated solids must progress downwards through the media to eventually break through.

DE filtration does not have this type breakthrough potential. As the cake thickness builds up, solids removal continually will improve as the run progresses. Breakthrough can occur only when the force applied to the flow is halted, interrupted, or is subjected to surge that may disturb the integrity of the cake. Any disturbance of flow should terminate the DE filter run to avoid the breakthrough potential.

Figure 3-6 shows the turbidity and particle count pattern for a pilot scale pressure DE filter operating at a higher application rate of 3.5 gpm/sq. ft. The gradual decrease of both turbidity and particle count is evident as the run progresses. For this filter operation, particle count in the 3- to 15-mm range was about 10 at the end of the run. Another significant difference between the two types of filters is that it may take an hour or two for the particle count to stabilize at the start of a DE filter run. This may be compared to that of a well-run rapid sand filter where initial breakthrough should stabilize in less than 30 minutes. The initial maximum level of rapid sand filter breakthrough is, however, higher than that for the DE filter operation. Again, most of the initial breakthrough in a DE filter is the fine, inert DE matter that has not stabilized in the precoat bridging, not the raw water particles.

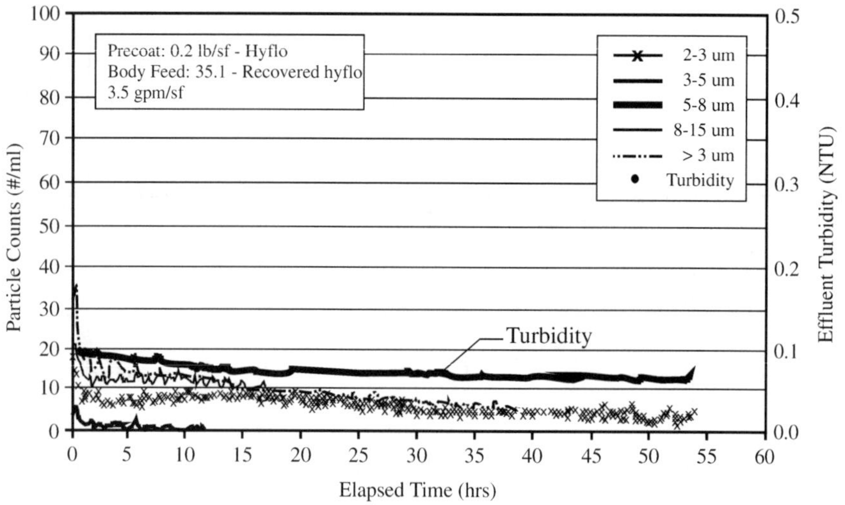

FIGURE 3-6. Particle Count vs. Elapsed Time

Color Removal

Color is present in source waters in colloidal or dissolved form. Color is measured in color unites (CUs) where 1.0 unit of color is that produced by 1.0 mg/l of platinum in distilled water. Color is expressed both as apparent and true color:

— Apparent color is generated by both dissolved and suspended color matter.
— True color is caused only by dissolved matter, determined from a filtered sample.

Color in source waters may result from a number of contaminants, the principal of which include soluble and suspended organics and iron. Suspended color may be removed in DE filtration as long as the colloidal and other fine particulate matter is charge-free. Oxidation with ozone, chlorine, and possibly potassium permanganate will have multiple effect on certain color constituents. Oxidation will both precipitate dissolved matter like the compounds of iron and manganese and literally bleach other matter by generating new compounds.

Figures 3-7 and 3-8 show the benefit of ozonating a prechlorinated supply. Both Figures show the unfiltered color of a prechlorinated supply after 16 hours of detention in a transmission main. In other words, some bleaching action of the reservoir source has occurred. As indicated in Figure 3-7, DE filtration reduced the source water apparent color of from 11 to 39 CU to levels of 1.0 to 9.0 CU or about an average of 5.3 CU. With ozonation (Figure 3-8) in a period where inlet water color ranged from 13.0 to 59.0 CU, overall color has been reduced to 5.0 CU or less except for the period of color surges. All color is apparent. These data were from a research program using full-scale DE filters with varying rates of preozonation. The few points of color above 5.0 CU could have been reduced further with greater rates of ozone addition.

Figure 3-7 illustrates how DE filtration will reduce suspended color matter. Any chlorine bleaching of dissolved color had already been completed before filtration. Figure 3-8 shows how ozone, the strongest of oxidants, can reduce color further by completing color precipitation and particle neutralization as well as with a bleaching action. For the water treatment illustrated, the predominant color present was contributed by organic matter with a slight iron content. Suspended color was essentially

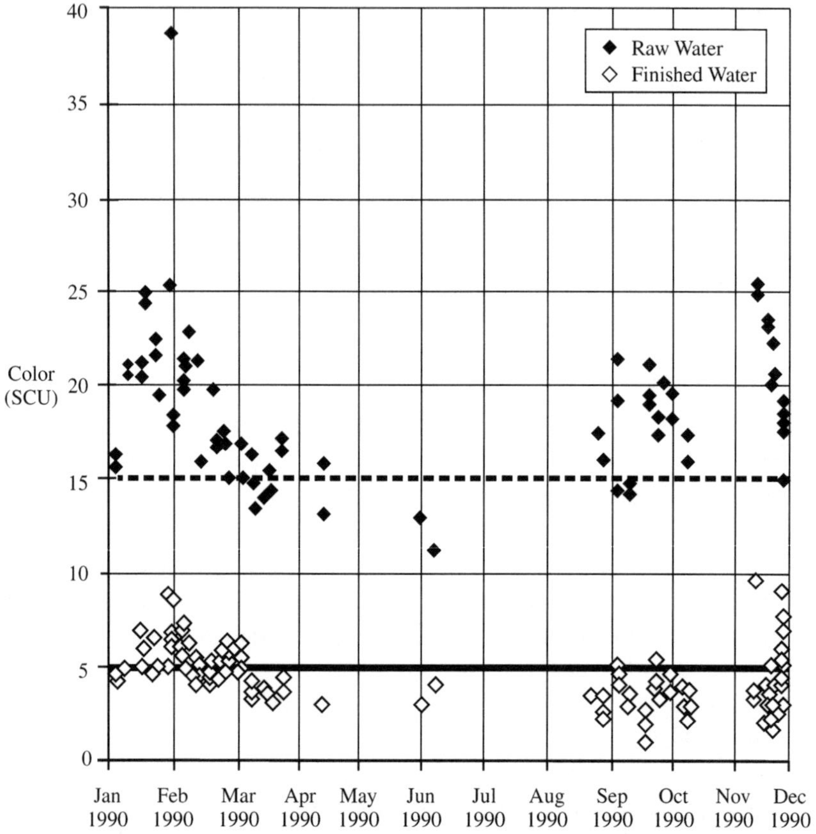

FIGURE 3-7. Color Removal (Without Ozone)—Annual Variations

colloidal in size. The prechlorination partially neutralized the colloidal color, which was carried further by preozonation to enable removal by DE filtration.

Iron and Manganese

The presence of iron and manganese in source waters exceeding the secondary limits of 0.30 mm and 0.05 mm respectively, will also inhibit DE filter operation. This situation is especially aggravated when pretreatment includes oxidation by ozone, chlorine, or potassium permanganate. Oxidized iron and manganese tend to coat or scale, which can entrap particles

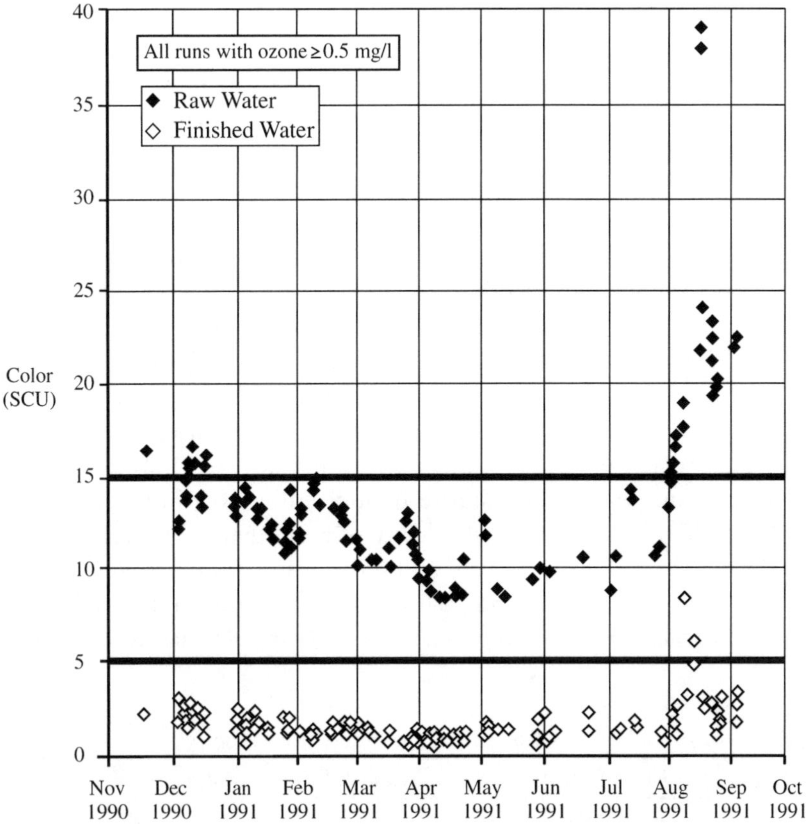

FIGURE 3-8. Color Removal (with Ozone)—Annual Variations

of DE on the surface and in the web of the woven wire or plastic septum. In rapid sand filtration, the oxidized soluble metals will scale or coat the granular media. Relatively soft in consistency, this scale is removed when the granular filter is backwashed. In addition, the time of travel through the media bed of 30 to 60 inches deep is usually sufficient to accomplish the scaling action before the water flow exits the filter. With DE cake thickness measured in fractions of an inch at the start of a run, the scaling action in the cake cannot be completed before the flow reaches the surface of the filter leaf.

Since both oxidized iron and oxidized manganese will readily scale or coat diatomaceous earth particles, removal can be accomplished in a contact basin where the body feed is mixed with the oxidized material. With sufficient detention, usually 10 minutes, the scaling will be fully completed before the flow is routed to the DE filter. Small in size, DE particles have a large aggregate surface area. Oxidation may be accomplished both by using dry chemicals mixed in solution and by oxidation with gaseous chlorine or ozone. The two methods are described later with some typical performance examples. Either method will require 10 minutes' detention for the oxidized material to coat the body feed. Iron can be precipitated as discrete particles or may be adsorbed to other particles. Depending on magnitude of flow, detention may be provided in vertical upflow tanks or in horizontal flow basins. It is important to minimize settling of the DE particles in the detention basin. Table 3.4 shows the particle size distribution and the settling velocity of each size particle found in Hyflo, a DE media commonly used in drinking water treatment. These and similar velocity determinations for other media may be used in the design of both vertical and horizontal basins to inhibit settling. For instance, an upflow basin should have a vertical velocity of more than 0.06 ft/sec to prevent the settling of DE particles of 250 microns in size. There will be particles of greater size present, but the design velocity for a straight-side tank would usually be much higher so that negligible settling would occur.

TABLE 3.4. Particle Size Distribution and Settling Velocity

% Finer	Mass Fraction	Micron Size	Settling Velocity (ft/s)
1	0.01	2.5	8.0E-06
4	0.03	3.5	0.000016
9	0.05	5	0.000032
24	0.15	8	0.000082
43	0.19	12	0.000184
55	0.12	17	0.000369
74	0.19	25	0.000795
87	0.13	40	0.00202
92	0.05	60	0.004491
94	0.02	90	0.009884
95	0.01	125	0.018489
99	0.04	175	0.034443
100	0.01	250	0.06443
Total	1.00		

Note: For Celite Hyflo DE Media

Chemical Oxidation

Dry chemicals used for chemical oxidation are as follows:

— Magnesium Oxide (MgO) is used to precipitate ferrous iron (Fe^3) to the substantially insoluble ferric hydrate [$Fe(OH)_3$].
— Potassium Permanganate ($KMnO_4$) is used to precipitate soluble manganese to the insoluble manganese oxide (MnO) precipitate.

Magnesium oxide or magnetite will also reduce a small amount of the manganese content. Similarly, potassium permanganate will reduce a small amount of the iron content. Table 3.5 shows these reductions in a two-stage plant. With high iron and high manganese, it is necessary to use two complete systems in series. The data shown are for treatment of a well supply in Milford, Mass., where both the iron and manganese content are high. In two-stage operation, removal of iron should precede manganese treatment. As indicated, the Milford operation reduced iron content of more than 3 mg/l and manganese of almost 1 mg/l to well below the recommended levels. As indicated, the magnesium oxide and potassium permanganate reduced the major portion of iron and manganese respectively. Where iron or manganese predominates in content, pilot tests should be conducted to determine whether a single stage, with MgO or $KMnO_{4,}$ will be appropriate to successfully achieve reduction of both metals.

Table 3.6 shows the results of jar tests prior to pilot runs to determine the effectiveness of manganese removal using $KMnO_4$ and DE. The jar test provides excellent simulation of a detention basin. As indicated in the Table, in these tests for a well supply in Norwalk, Conn., practically complete removal of manganese was achieved.

Special care must be exercised in the use of potassium permanganate to avoid producing "pink" water. Just a slight amount of $KMnO_4$ above the demand can produce such color. While not harmful, it is not a color appreciated by the consumer.

Ozonation

Ozone is the strongest oxidant available for water treatment. Because of its strong oxidation properties, ozone demand may be high in certain source waters because of type and amount of dissolved solids that may be present. Tests with northeastern groundwaters where hardness is moder-

TABLE 3.5. Iron and Manganese Removal: Two-Stage Process

Date		Raw Water Plant Tap	Filtrate #1 (mg/l)	Filtrate #2 (mg/l)	% Reduction
9/27	Tot. Fe	3.24	0.31	0.0	100
	Fe^{11}	2.46	0.06	0.0	100
	Mn	0.8	0.6	0.011	98.6
9/28	Total Fe	3.33	0.39	0.0	100
	Fe^{11}	2.51	0.06	0.0	100
	Mn	0.8	0.65	0.017	97.9
9/29	Total Fe	3.41	0.28	0.0	100
	Fe^{11}	2.48	0.04	0.0	100
	Mn	0.8	0.65	0.013	98.4
9/30	Total Fe	3.29	0.28	0.0	100
	Fe^{11}	2.38	0.05	0.0	100
	Mn	0.8	0.65	0.018	97.6
10/1	Total Fe	3.24	0.26	0.0	100
	Fe^{11}	2.28	0.07	0.0	100
	Mn	0.8	0.6	0.019	97.6
10/3	Total Fe	3.28	0.24	0.0	100
	Fe^{11}	2.21	0.05	0.0	100
	Mn	0.8	0.6	00.18	97.8

Note: For all runs above: Average Flow = 1.0 mgd; DE Body Feed = 50 to 60 mgd/l; MgO = 10 to 14 mg/l; $KMnO_4$ = 1 to 2 mg/l

ately high indicate that rather low ozonation rates would be sufficient to condition iron and manganese for removal. Western waters, much higher in mineral content, will sometimes require ozonation rates as high as 2.0 mg/l for iron and 4.0 mg/l for manganese. The demand for the overall dissolved solids content controls the total amount of ozone required to effectively remove iron and manganese.

Because of significant facility costs, ozone would not be the oxidant of choice for just iron and manganese treatment. Where ozone may be required for other purposes like disinfection or conditioning of suspended organic matter, iron and/or manganese treatment may be effectively "piggybacked" for cost savings.

While less data is available for ozonated supplies, it appears that after ozonation, 10 minutes' detention for contact with the DE filter body feed

TABLE 3.6. Well Water Supply: DE Removal of Manganese

Test No.	$KMnO_4$ (mg/L)	DE (mg/L)	Manganese (mg/L)	
			Raw Water	Treated
Jar Tests				
1	0.9	50	0.78	0.04
2	0.5	55	0.78	0.15
3	0.9	50	0.72	0.04
Pilot Tests				
1	0.5	50	0.29	0.02
2	1.0	50	0.68	0.05
3	1.0	50	0.77	0.02
4	1.0	50	0.62	0.03
5	1.0	50	0.60	0.003
6	1.0	50	0.67	0.01

Note: Contact basin detention = 10 minutes.

would be sufficient. Detention time requirements in ozonation will also vary. Discussion of ozone feed rates and detention time are beyond the scope of this text and should be determined in pilot tests.

Other Options

Many well supplies pass through air strippers to remove volatile contaminants. While an air stripper would not be provided just to oxidize iron or manganese, some conversion of the soluble metals is accomplished in this action. Where a slight reduction of iron or manganese is required, the oxidation in the stripping tower may be sufficient followed by 10 minutes' detention with the body feed. In any case, the aeration provided will partially reduce the requirements for chemical oxidation (MgO and $KMnO_4$) that would follow. The benefits of the air stripping should be determined in pilot tests.

Chlorine is another strong oxidant that might be used to convert soluble iron and manganese. Unfortunately, if dissolved organics are present in a well water or surface supply in sufficient amounts, disinfection byproducts may be formed in the reaction with chlorine. There are many well supplies, however, where the formation of disinfection byproducts will be negligible. Both ozone and chlorine used for oxidizing soluble iron and manganese will accomplish partial to complete disinfection, depending on feed rates and detention time. In any case, preoxidation from aeration through ozonation will reduce the chlorine demand and the chlorine rate required to maintain a residual in the distribution system.

Disinfection Byproducts

Disinfection byproducts formed in water treatment by the reaction of certain NOM in disinfection, principally with chlorine, can be reduced in water treatment by removal of a sufficient amount of the organic precursors. The principal organic removal mechanism in rapid sand filtration is coagulation/filtration. Because of the nature of the less permeable DE filter media, this type of removal is impossible. The use of BAC media preceded by ozonation has been proven to be an effective means for not only reducing the organic content, but also for improving the efficiency of DE filtration itself. It is noted that BAC should be preceded by ozonation to convert the less biodegradable organic matter to the more biodegradable forms. As mentioned before, ozonation is effective in a DE filtration train to neutralize negatively charged colloidal matter to improve removal. BAC contact basins would follow ozonation and precede the DE filters. With this arrangement, the multiple benefits provided include:

1. Neutralization of negatively-charged particles (ozonation)
2. Removal of larger-sized particles in the BAC contactor by entrapment
3. Reduction of biodegradable organics to both reduce the formation of disinfection byproducts and inhibit regrowth of pathogenic organisms in the distribution system.

Biological Activated Carbon Process

Particle neutralization and removal by entrapment have been discussed previously. In BAC treatment of drinking water, heterotrophic organisms that are practically ubiquitous in surface waters obtain nourishment from the ingestion of organic matter. More specifically, heterotrophic bacteria colonized on granular activated carbon (GAC) media in a supplemental contactor feed and metabolize the biodegradable natural organic matter in the water to produce essentially non-biodegradable compounds. BAC in water treatment is simply nature's process that is accelerated by providing an optimized environment for heterotrophic activity. The breakdown of organics by heterotrophic activity occurs in most lakes, reservoirs, and river systems. Heterotrophic bacteria are active in the overlying water, on the bottom surface, and into the uncompacted zone of bottom sediments in these water systems. GAC media provides a large surface area on which the heterotrophic bacteria colonize, proliferate, and eventually reach a state of equilibrium in progressive stages. It has been estimated that a cubic centimeter of GAC media may support between 50 and 150 million bacteria compared to 2 to 5 million bacteria in a similar volume of sand.

After passing through the life cycle to die off, the expended bacteria continually slough off the media and enter the effluent stream. When the available head loss is expended due to both trapped source water solids and accumulation of trapped expended bacteria, the contactor must be backwashed. In essence, the BAC contactor in DE filtration application may be considered to be a roughing filter. While it operates in a manner similar to the rapid sand filter, no coagulants are used and backwash frequencies may be a matter of weeks rather than days.

Typical Performance

The performance features reviewed later are from operation of an ozone-BAC-DE filtration train treating water from a reservoir impoundment that is moderately eutrophic. Significant features of the BAC column and operation are as follows:

— GAC average media size: 1.4 mm
— Media depth: 60 inches
— Flow application: 4.5 gpm/sq. ft.
— Detention: 10 minutes
— Ozonation rates: 1.0 to 1.5 mg/l.

Table 3.7 shows the effectiveness of particle removal through the contactor. Removals varied from more than 90% in June to more than 80% in April. Data is not available, but it would be expected that removals would be less in the January to March period of coldest water temperature. Temperature affects removals in two ways. First, particle separation is less efficient with cold, more viscous water flow. Second, bacteria activity is less prolific in colder temperatures, which would tend to result in a more permeable media. Of course, seasonal reservoir turnover and rainfall-runoff activity any time of the year may result in generation of suspended matter with different separation characteristics.

The effectiveness of organic or dissolved organic carbon (DOC) reduction in a BAC contactor is shown in Tables 3.8 and 3.9. Table 3.8 shows the reduction in DOC as well as BDOC. DOC reductions from raw water through ozonation and BAC contact are shown in the first column. Percent removals are for each successive step. For instance, on June 3rd, reduction of raw water DOC through ozonation was 4%. Reduction from the ozonated water through BAC was 16% with only 1% reduction of the DOC in the BAC effluent through DE filtration. Performances for the three dates shown are similar with the major reduction occurring in the BAC contactor treating the ozonated water. Reductions in the second column for BDOC are even more dramatic. As indicated, BDOC rises as non-biodegradable organics are converted to the biodegradable form through ozonation. Reductions of BDOC through the BAC column are from 31 to 41% compared to the 16 to18 % reduction of DOC.

Table 3.9 shows the effectiveness of THM and HAA5 reduction through the BAC process. Actually, most reduction occurs with ozona-

TABLE 3.7. BAC Contact Basin: Particle Reduction

		Particle Counts (μm)				
Date		(2–3)	(3–6)	(5–8)	(8–15)	(>15)
4/18/97	Raw	1286	3995	1298	538	80
	BAC	493	1089	232	77	9
4/24/97	Raw	1321	3606	1365	632	120
	BAC	130	277	100	43	6
5/1/97	Raw	1402	3708	1305	614	98
	BAC	61	119	36	43	4
5/15/97	Raw	1289	3234	1275	546	76
	BAC	183	350	97	44	5
6/5/97	Raw	1072	2740	1413	951	203
	BAC	127	278	89	29	8

tion. The BAC column, therefore, reduces the BDOC content, which inhibits regrowth potential in the distribution system. While it is not simple to define the results in a quantative manner, comparative tests showed that the ozone-BAC-DE filtration trains produced water as stable or more stable than that from rapid sand filter trains.

For treatment of the water illustrated previously, the ozone-BAC-DE filtration train was effective in reducing the level of the two major disinfection byproducts of concern to levels below the maximum expected for the second stage EPA requirements. Moreover, the process produced a relatively stable water to inhibit distribution system regrowth not controlled by a disinfection residual.

TABLE 3.8. BAC Contact Basin: DOC Reduction

		DOC		BDOC	
DATE		MG/L	% REMOVAL	MG/L	% REMOVAL
5/20/97	Raw	3.01		0.45	
	After O_3	2.84	6	0.96	—
	BAC	2.33	18	0.57	41
6/03/97	Raw	2.87		0.63	
	After O_3	2.75	4	1.16	—
	BAC	2.30	16	0.75	35
	DE	2.28	1	0.70	7
7/01/97	Raw	2.52		0.57	
	After O_3	2.44	3	1.21	—
	BAC	2.03	17	0.84	61
	DE	2.00	1	0.84	—

TABLE 3.9. DBP Formation: Ozone–BAC/DE Filtration

LOCATION	SDS THMs (μG/L)	SDS HHA5 (μG/L)
Raw Water	86.3	62
After O_3	46.2	32
After BAC	32.9	23
DE Filter Effluent	34.4	24

4
Filter and Ancillary Systems

Several hydraulic, material handling, and equipment systems are required to complete an operating DE filter installation, the principal ones of which include the following:

— *Basic Filter System:* equipment, piping, and controls for the precoat, filtering, and spent cake sluicing functions of the filter. These may include pumps to deliver raw water, slurry feed pumps for precoat and body feed, pressure and flow controls for filtered water, as well as tanks and piping for precoat, precoat recirculation, body feed, and the discharge of sluiced cake.

— *DE Delivery, Storage, and Preparation:* equipment for unloading bagged DE delivered by truck and for moving between places for storage and preparation. Tanks, mixers, and other equipment required for emptying bagged DE in the preparation for use as precoat and body feed in slurry form.

— *Sluice Slurry Systems:* tanks, piping, and other equipment required to convey slurry sluiced from the filters to a place of storage and preparation for ultimate disposal.

The extent of the above systems will vary according to the requirements for each particular installation. Variations will be influenced by the quality of the raw water supply, system capacity, site features, and the distribution system delivery needs.

The basic filter system features will differ for pressure type and vacuum type DE filters. DE delivery handling, preparation, and disposal features, however, would be essentially the same for both type filters.

Basic Pressure Filter System

The elements of the basic system serving the DE filter may be grouped into three parts in accordance with the three operating steps as discussed later and shown in Figure 4-1.

Precoat

As indicated in Figure 4-1, the basic elements include a recycle pump, recirculation tank, and recirculation lines, in addition to the precoat slurry tank and feeder. In the more conventional systems, the empty filter, recirculation tank, and pipelines, are first filled with water to a static level just above the filter shell. As the pump recirculates the flow around the filter-tank loop, a premeasured amount of virgin DE is added in slurry form by pump to the flow entering the filter. Often the filter feed pump is used to recirculate the precoat flow. In other situations, it has been found advisable to use a separate pump for this purpose. All precoat tanks must be fitted with agitator devices to keep the diatomite slurry in suspension. Accommodations should be provided for charging the precoat tank with a measured amount of diatomaceous earth. The tank bottom should be located at an elevation that ensures a positive pressure in the pump suction. Precoat feed piping should be arranged to prevent back siphonage of flow from filter to tank. Precoat tanks should have dished, coned, or slanted bottoms for ease of draining the tank for cleaning. Other important features of design include:

— Precoat tank agitators and pump suction should not create vortexing, which can cause air entrainment. Tank baffles should be installed if necessary to prevent vortexing.
— The return recirculation line should enter the recirculation tank at an elevation below the minimum liquid level to avoid creating turbulence and air entrainment.
— Recirculation pump suction seals must be sized and maintained to inhibit air or gas inlet.

Sizing of tanks depends on the recirculation detail. In the more conventional, more commonly used design, the recirculation tank is sized so that the total precoat system liquid volume would be about 125% of the filter tank and connecting piping volume. Since the piping volume is relatively small, the recirculation tank capacity should be from 20 to 25% of the filter shell.

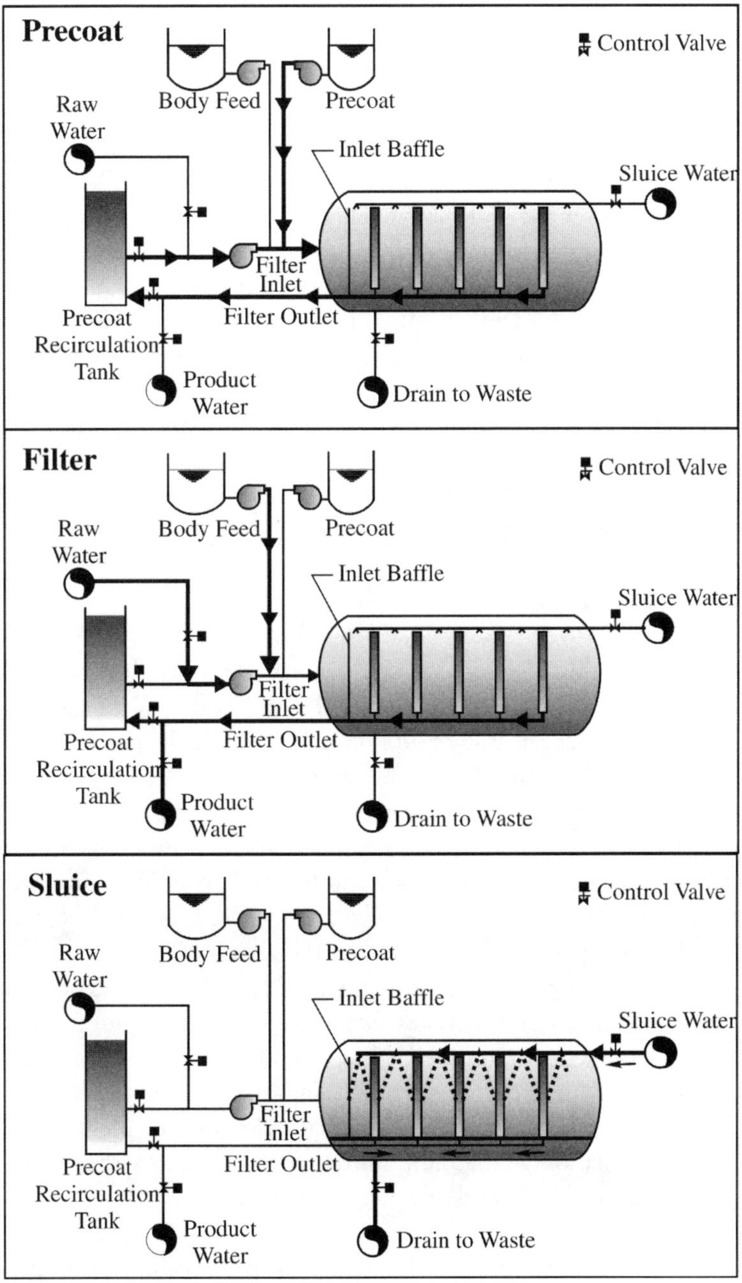

FIGURE 4-1. Typical Pressure Filter System Operations

In an alternative design, the recirculation tank is eliminated and the filter and piping may be filled directly from a pressure water supply. Where such supply is potable water, an approved backflow prevention device must be installed in the pressure connection. With this type arrangement, air is vented from the filter shell through an air/vacuum release valve. Just before the filter shell is completely filled, the precoat slurry is introduced and the recirculation initiated. Referring to the Figure, the recirculation tank with its inlet and outlet valves are eliminated and the pump suction-filter outlet lines are tied across with a single valve in the place of the tank. While this alternative arrangement is simple to operate and less costly, it does eliminate a convenient means for the operator to observe when the recirculation system is full, the precoat is complete, and the recirculation flow runs clear. Sight ports can be installed in the filter shell and the recirculation line to permit observation of flow clarity. Where cost and installation space are not major issues, the recirculation tank arrangement would be preferred.

Filter Cycle/Body Feed

Once precoating is complete, raw water is introduced to the filter pump suction, and the recirculating tank inlet and outlet valves are closed in a simultaneous action. Body feed is initiated. While the precoat system provides for batch application of diatomite, the body feed system must feed at a continuous, metered rate. This means that slurry in the body feed tank must be mixed in reasonably accurate concentrations and must be fed to the filter inlet flow at a controlled rate in proportion to the flow. Slurry concentrations can be as high as 18 percent, but are best kept in the 5 to 12% range to minimize meter pump abrasion.

Where flow to the filter is maintained at a constant rate, the metering pump may be set at a constant rate of feed. Metering pumps may be the piston/diaphragm or the peristaltic type. Pumps should be suitable for abrasive slurry service. Single outlet diaphragm pump discharges may be located at the bottom of the diaphragm chamber to prevent separation and accumulation of diatomite in the chamber. Other special body feed system features that should be considered include:

— Slurry tanks should be fitted with low speed turbine or propeller type agitators (40–60 rpm) to maintain the diatomite in a relatively evenly distributed suspension.

— Metering pump control design should permit change in the setting (rate) any time while it is in operation.
— Where flow to the DE filter can vary, the metering pump should be automatically paced by the inlet flow meter controls.

Figure 4-1 indicates the elements of the system that are in service for the filtering cycle discussed previously.

Sluice Water

When the accumulation of filter cake generates maximum head, the filter run is terminated and the cake is sluiced from the leaves. As shown in Figure 4-1, in the sluicing cycle, all filter valves are closed except for the sluiced slurry drain valve. Sluice water, at pressures more than 60 psi, is introduced through spray nozzles directed at the filter leaves. Except for the smaller vertical type filters fitted with fixed nozzles, spray nozzles are usually mounted on a manifold header or headers that rotate to give full spray coverage to each side of the filter leaf. Oscillating mechanisms are mounted outside the filter shell and the header passing through to the interior of the shell may be provided with stuffing boxes or, more frequently, with mechanical seals to withstand the filter operating pressure. As discussed in the previous chapter, in the rotating filter leaf design, the cake is sluiced by fixed sprays as the leaves turn on the center manifold. Pressure filters must be fitted with vacuum-air relief valves to permit uninhibited draining as well as filling of the filter shell.

Depending on the size and configuration of the DE filter, supplementary sprays may be necessary to move the sluiced slurry to the exit drain. The sluiced slurry is conveyed to a holding tank before preparation for ultimate disposal.

System Control Elements

Every DE pressure system is fitted with a flow control that consists of a flow measuring device and associated instrumentation to actuate a throttling valve or to regulate the speed of a variable speed pump. The more common flow element is a venturi insert. For smaller capacity systems, orifices or even propeller meters would be appropriate. The flow measuring device and associated transmitter would control a preset fixed flow through the filter and could pace the body feed pump. The same signal would also control the precoat recirculation rate. While pressure type DE filters may be operated at leaf application rates of more than 3.0 gpm/sq. ft. in the filter cycle, precoat is applied at rates of between 0.5 to 1.0 gpm/sq. ft. The other principal measurement is the loss of head, the difference in pressure between the filter inlet and the outlet product water lines.

The operation of the diatomaceous earth filter may be controlled by means of an automatic computer program or by programmable controllers. As shown on the flow diagram, all valves would be electric motor operated, the most common type used in drinking water filter installations. For manual control of the three DE filter steps, the valves could be actuated by open-close pushbuttons. For computer operation, valve control might be as shown in Table 4.1, starting with precoat and an empty filter. The flow paths for each filtration step are as shown in Figure 4-1. Only the smaller sized installations would employ handwheel-operated valves for the process functions.

As indicated in the Table, the time for the precoat and sluicing cycle as well as the time between all cycles may be controlled by a timer. As conditions or treatment requirements change, the time periods should be adjusted. For instance, a longer cycle would be required for sluicing greater cake thickness generated by an increase in body feed.

In many situations, it would be advantageous to use a combination manual-automatic control. The operator starts each of the three cycles individually by pushbutton, which actuates all valve and equipment operation automatically as established by the computer program. The advantage of this mode is that the operator may inspect the completeness of precoat and sluicing before proceeding with the next step. The filtering cycle would always be stopped automatically by the established maximum loss of head.

While the DE filter operation is not usually prone to breakthrough of solids, operation must be protected against loss of inlet flow, which can result in loss of integrity of the collected cake. When flow is halted or flow drops below a minimum rate of 0.25 gpm/sq. ft., the filter run must be terminated and the filter sluiced to prepare for the next run.

Basic Vacuum Filter System

As shown in Figure 4-2, the vacuum filter is operated in the same manner as the pressure filter in three steps. Operation of the vacuum filter system is identical to that of the pressure type except for the following:

TABLE 4.1. Control of DE Filter Operation

1. Fill the system with filtered water to the level set by a control in the precoat recirculation tank.
2. Activated by the same level control, start the recirculation pump to maintain the flow rate set by the flow control through the path established by previously opened flow valves shown on the Figure for precoat.
3. Add the premeasured virgin DE slurry by pump to the recirculated flow from the precoat tank.
4. Continue the recirculation until the precoat is completed, indicated by the clarity of the recirculated flow. Time of recirculation may be determined by a turbidimeter taking sample from the recirculation line or by allowing sufficient time (set on a timer device) for the precoat to be completed.
5. When the precoat is completed, the flow path is switched to that shown on the Figure for filter by opening and closing the valves indicated in simultaneous operation along with the following:
   - Stop the recirculation pump and start the filter feed pump (if different units) or increase flow rate to that set on the flow control for filtering.
   - Start the body feed metering pump paced by flow control.
6. When the filtering cycle is completed, switch to the sluice cycle.
   - Stop the filter inlet feed (valve and/or pump).
   - Stop body feed.
   - Close all valves except those for the sluice water inlet flow and drain to waste.
7. Continue the sluicing operation for the period set on a timer; then shut sluice water and drain valves.
8. Either immediately after or after a "wait" period set on a timer to allow inspection of system status, open and close valves to establish the flow path for precoat.
9. Start cycles over with Step 1.

Precoat

Since the vacuum filter is open to the atmosphere, no precoat recirculation tank is required. Precoating may be observed directly in the open filter.

Filter

The filter flow rate is monitored at the discharge of the pump taking suction from the vacuum filter. The water level in the filter is maintained by

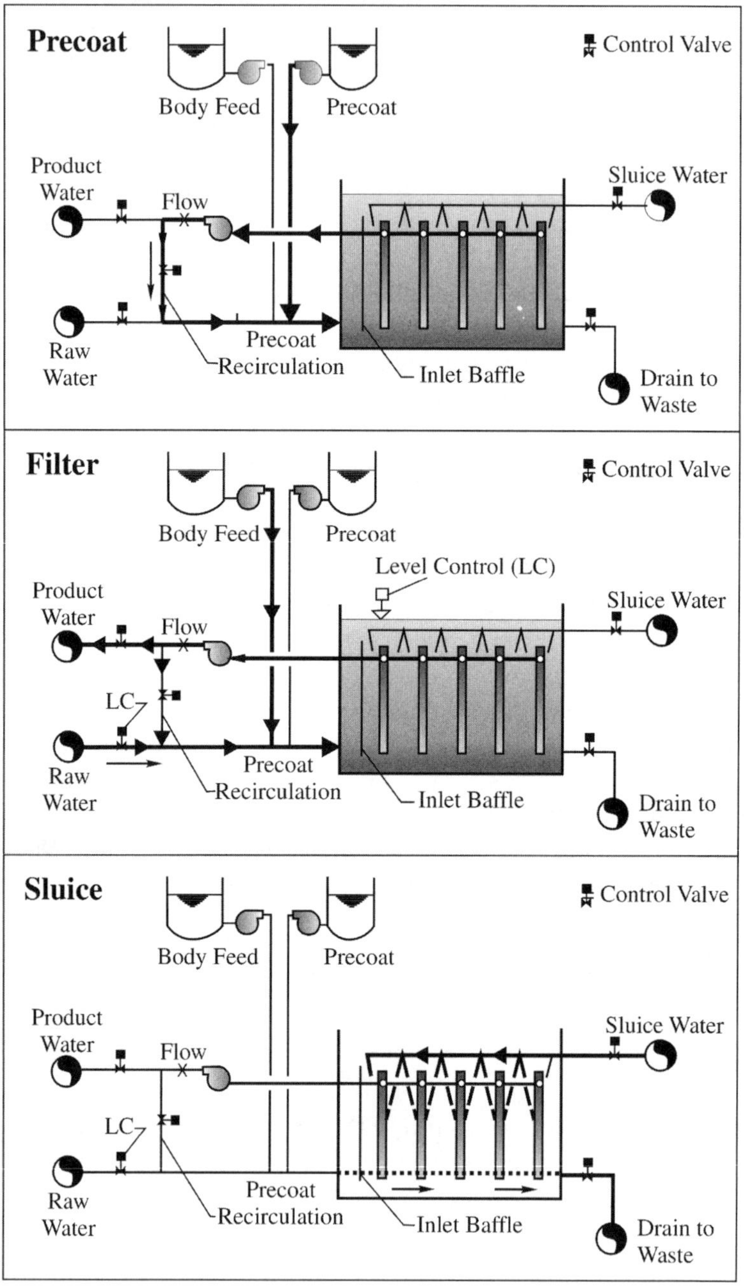

FIGURE 4-2. Typical Vacuum Filter System Operations

means of a level control that activates the inlet flow valve as indicated in the figure. For smaller capacity units, water level in the filter may be maintained by means of a simple float-operated valve in the inlet line. In this case, the inlet valve would be immediately adjacent to the filter with the float in the filter basin.

Sluicing

With the vacuum filter open to the atmosphere, the filter leaves may be sluiced manually with a high-pressure hose. This is a positive method of cleaning where the surfaces of the leaves are visible. The operator can clearly see when the leaves are clean and the cake residue is flushed down the drain. Rotating filter sprays may also be used for sluicing vacuum filters, but the unit must then be provided with a water-tight cover which loses some of the advantages of visibility for sluicing and the other steps in the filtration cycle.

Controls

Controls for the vacuum type filter would be identical to those of the pressure alternative in concept. The principal differences are the lower operating pressures and the level control described previously to maintain the level in the filter. In pressure operation, the level control is used only for filling the filter system for precoating. In vacuum filter operation, the level control is mounted to measure and control the level in the filter itself for both the precoat and filtering cycles. For precoat, the level control actuates the inlet of filtered water to fill the system. For filtering, the level control actuates the inlet control valve to maintain constant level through the filtering cycle. As mentioned, instead of actuating motor operated valves, the level control may be strictly mechanical, employing float operated valves.

While full computer program operation is possible in vacuum filter operation, its benefits are doubtful. The particular advantage of the vacuum filter is its open operation where the effectiveness of the three cycles of operation may be clearly observed. Computer control would rarely be considered for the smaller capacity vacuum filter system. For systems of 1.0 mgd and greater capacity, motor-operated valves should be considered. Where a limited operations staff exists, the modified manual-automatic controls of filter operation may be used effectively.

With modified manual-automatic control of vacuum filter operation, each cycle is initiated by pushbutton. The computer program controls actuation of the valves to establish the required flow paths. The staff operator/observer is responsible for the following:

1. Introducing precoat material to the recirculated flow by initiating the feed device
2. Halting precoat recirculation when sufficient clarity is established in the recirculated flow to switch to the filtering cycle
3. Halting the filtering cycle when maximum head loss is reached (indicated by automatic alarm)
4. Starting the sluicing operation, either manually with high-pressure hose or by initiating rotating sprays (covered units only)
5. Halting sluicing when observed to be complete.

Multiple Filter Arrangements

The general piping arrangements for the two basic types of DE filters were discussed previously and shown in Figures 4-1 and 4-2 for individual filters. Economy of installation and simplification of operation may be achieved by arranging pairs of filters for parallel operation and by providing banks of filters with common precoat facilities and common body feed storage. These two optional arrangements are shown in Figures 4-3 and 4-4 and discussed later. While the piping arrangements in the Figures are for the pressure type filters, the concepts are also appropriate for multiple vacuum filter units.

Filter Pairs

As shown in Figure 4-3, except for sluice drain to waste, the piping for the two filters is interconnected to permit a single inlet, outlet, sluice water, recirculation tank, body feed, and precoat for the pair. In operation, the pair of filters function as a single unit identical to that shown in Figure 4-1 and Figure 4-2 for single unit. The principal concern is that the piping system be designed so that the head loss for the recirculation (precoat) and filtering modes be identical from inlet to outlet for each unit. Expressed in terms of flow, the piping design should be such that the recirculation and filtering flow rates through each filter unit in the pair is identical. If, for instance, the piping run from inlet and outlet is longer for one of the filters, greater head loss would result for this unit in operation, causing uneven flow division between the units. The flow precoat thickness and filter cake accumulation would be greater in the filter with the

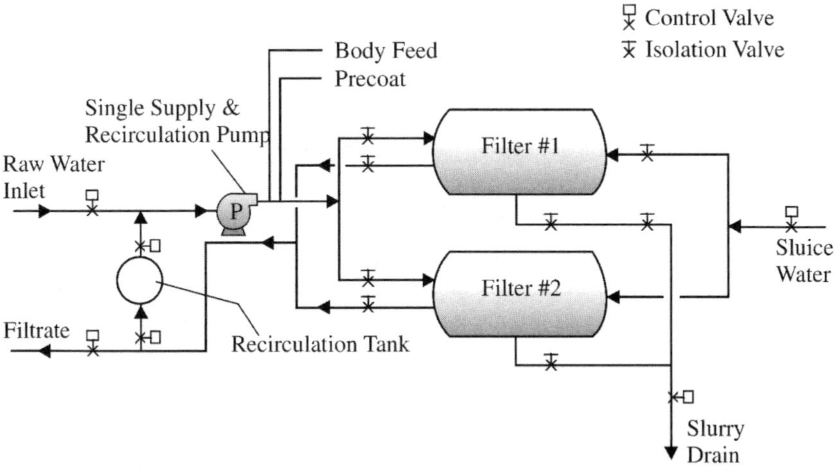

FIGURE 4-3. Dual Pressure Filters—Common Ancillary Systems

least head loss. Since no head loss can be predicted perfectly in design, stops may be provided on the inlet valves to each filter to allow adjusting flow evenly between the filters. The adjustments are made by setting the stops to provide the throttling required on the low head loss unit when operating the filters (clean) individually at a fixed flow rate.

Filter leaves may clog partially with time, which results in changing head loss conditions. It would be desirable, therefore, to adjust the division of flow periodically to ensure equal head loss and efficient operation of the pair of filters.

Arranging DE filters in pairs makes it possible to eliminate individual pump units for inlet flow, precoat, and body feed as well as tankage for precoat recirculation, precoat slurry, and body feed slurry. Duplicate controls for flow and head loss may also be eliminated. Use of individual rather than double ancillary facilities reduces installation, maintenance, and operation costs.

It would only be practical to consider the paired filter arrangement when six or more filters are required in an installation. In essence, a pair of filters become a single operating system; it is always desirable to have at least three filters in any installation to allow for a reasonably constant production rate and to minimize the impact of removing filters for maintenance. Another advantage of the paired filters is that with proper rate controls for inlet and body feed metering pumps, a single unit in the pair may be removed for maintenance while the other unit remains in operation.

Filter Batteries

Another way to save installation, operation, and maintenance costs is to arrange multiple filter units in parallel with common ancillary precoat and body feed tank facilities. Each filter would be operated as a separate unit, but there would be a single precoat recirculation system along with a single day tank to store body feed slurry. This arrangement is shown in Figure 4-4 for four filter units. The principal operating requirement for the battery arrangement is that filter sluicing be scheduled so there is a reasonable amount of time between the precoat operation for the filters in the battery. Since some sluicing and precoating cycles together can consume the best part of an hour, it is reasonable to separate sluicing and precoating for individual units for up to two hours.

Except for the precoating system and the body feed day storage, the other aspects of the multiple DE filter system arrangement would be the same as for individual filters. Each filter would have separate supply pumps and control systems.

Additional cost and operating efficiency benefits may be realized by arranging filter pairs in a battery arrangement. In this case, each of the

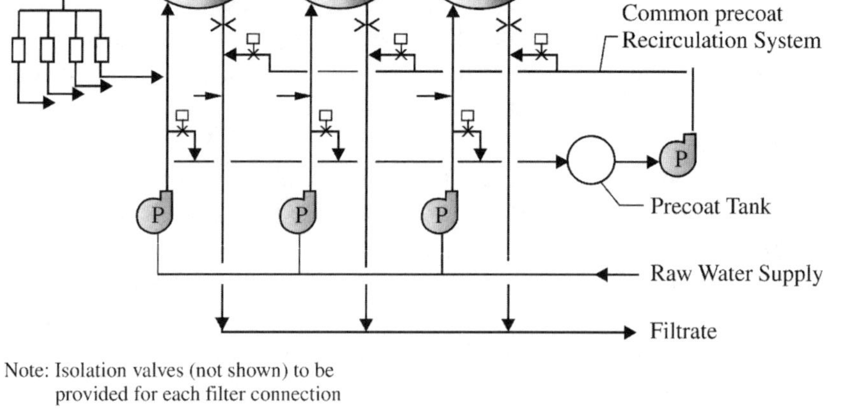

FIGURE 4-4. Filter Battery—Common Precoat and Body Feed Systems

individual filters in Figure 4-4 would be replaced by a pair of filters. The net result would be a reduction by half in the number of control systems, supply pumps, and body feed pumps, as well as a single precoat system.

Delivery and Storage of Dry Diatomaceous Earth

Diatomaceous earth is a powdered, essentially silicious material for which special precautions must be observed in handling. While not categorized a hazardous material, the basic face masks and eye glasses should be worn by operating personnel when unloading or transferring dry material.

Diatomaceous earth is available for delivery in:

- 50-lb paper, plastic, or plastic/paper bags
- Reusable woven plastic bags holding from 900 to 1,000 lbs
- Bulk, delivered in 15- to 20-ton tank-like trucks fitted with pneumatic discharge systems.

Bags are shipped on wood pallets. For 50-lb bags, there are three bags to the layer, nine layers high, for a total of 27 bags (1,250 lbs). The larger 900-1,000 lbs bulk bags are placed on pallets, two high. Special features for unloading, storing, and handling DE delivered in bags are discussed later.

Bulk DE delivery would be considered only for installations where unusually large amounts would be required. The cost effectiveness of such delivery would also be influenced by the delivery radius of the special purpose trucks used for conveying from the source. In the United States, bulk delivery would almost be limited to the western states. In addition, the high silos required to store the bulk DE would also limit its use to areas zoned for industrial development. Further discussions of the somewhat complicated unloading and conveying systems required for bulk DE are beyond the scope of this text.

50-lb Bags

Fifty pound bags stacked on wood pallets must be moved by forklift. Individual 50-lb bags may be picked up manually and carried for short distances. The type of forklift used depends on the magnitude of the pallets to be unloaded and the distances to be conveyed from truck to a place of storage, and then to the place of use. Forklifts may be the simple handpropelled jack type, the motorized walk-behind type, or the completely motorized unit on which the operator is seated.

As shown in Figure 4-5, the pallet is unloaded from the truck by forklift and moved to a place of storage. By using plywood sheets on top of the lower pallet stack, storage may be two stacks high. Only the motorized forklift is suitable for the two-pallet stack.

When needed, a pallet is moved to the place of use, usually to a bag breaker discharging to a slurry tank. The operator places individual bags in the hinged bag breaker door. The bag breaker is fitted with a dust collector to contain DE released in the breaker enclosure. For small plants, individual bags may be dumped directly, by hand, into the open tops of slurry tanks. Special precaution must be exercised with handling such as a well-ventilated environment. The enclosed bag breaker fitted with a dust collector is recommended and in many states may be required in the regulations.

The most common problem results from 50-lb bags breaking because of careless handling somewhere en route. All DE spills must be quickly picked up and washed down to protect the plant working environment. A wet-dry vacuum unit would simplify handling DE spills.

Large Woven or Bulk Bags

The design of the large woven or bulk bags is such that a ride-on forklift with a high lift is required for handling. As shown in Figure 4-6, each bulk bag has four top loops, positioned for the standard forklift. Two bags are stacked on a single wood pallet for shipping and storage. The double stacks are moved on the pallet from the delivery van to a place of storage in the treatment plant. The filled size of the bulk bag is approximately 47 in. × 47 in. × 47 in. When stacked on pallets, the bag tends to bulge. When lifted, the bag will be somewhat elongated.

When required, the forklift will lift individual bags to the place of use or the place where the DE is mixed in slurry form for use as precoat and body feed in process. In Figure 4-7, the bulk bag is positioned over the opening in a slurry tank. The operator then manually pulls down the bottom discharge spout and loosens the tie string to empty into the tank. In this method, the bag may be emptied at too fast a rate to permit uniform mixing. This problem is eliminated by attaching the discharge spout to the inlet of a volumetric feeder of the screw or roto-lock type as shown in Figure 4-8. The feeder empties the bag at a rate that ensures complete

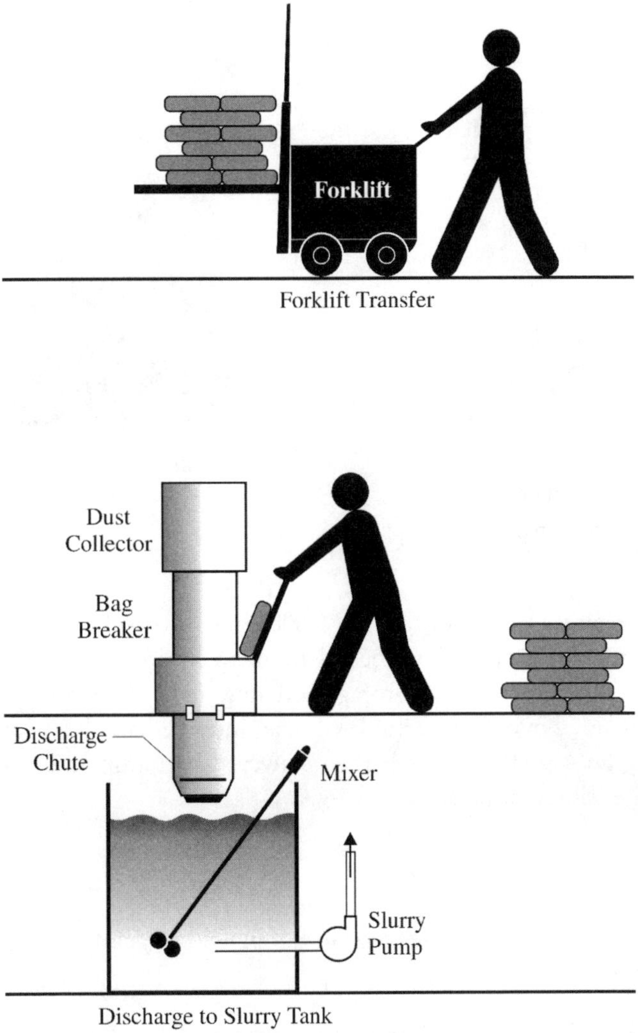

FIGURE 4-5. 50-lb. Bag Pallet Handling

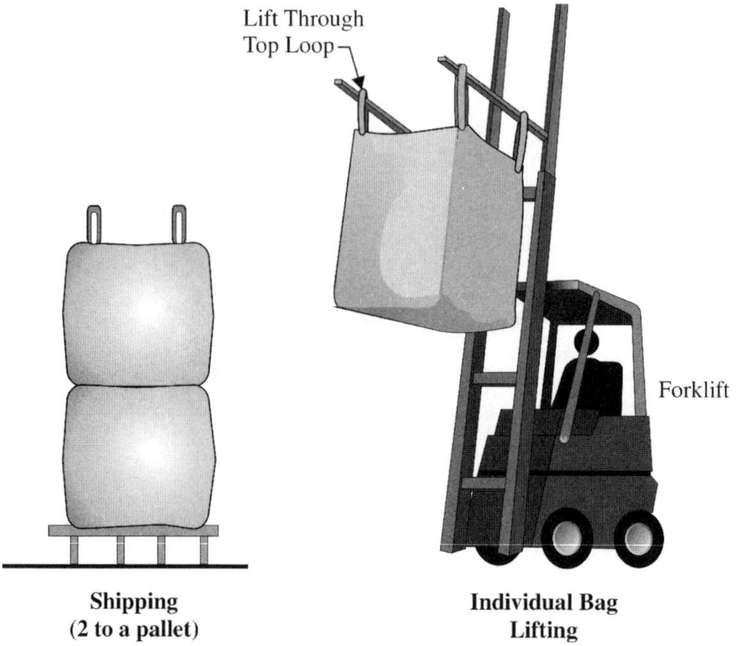

FIGURE 4-6. Bulk Bag Shipping Handling

mixing without settled clumps collecting on the bottom of the tank. Vibrators are mounted on the bag frame to ensure complete emptying.

If the place of slurry storage is remote from the place of mixing, the dry material may first be fluidized in an intermediate slurry tank as shown in Figure 4-9. The slurry is then conveyed by pump and pipeline to the place of slurry storage.

Selection

The type of dry DE delivery to adopt will depend on several factors, including:

- — Availability (delivery time) from source or distribution center
- — Amount of DE used on the average and for possible seasonal variations
- — Storage space.

Certainly, 50-lb bags would suffice for a small capacity plant using up to 200 lbs a day of diatomaceous earth. A 27-bag pallet would last more

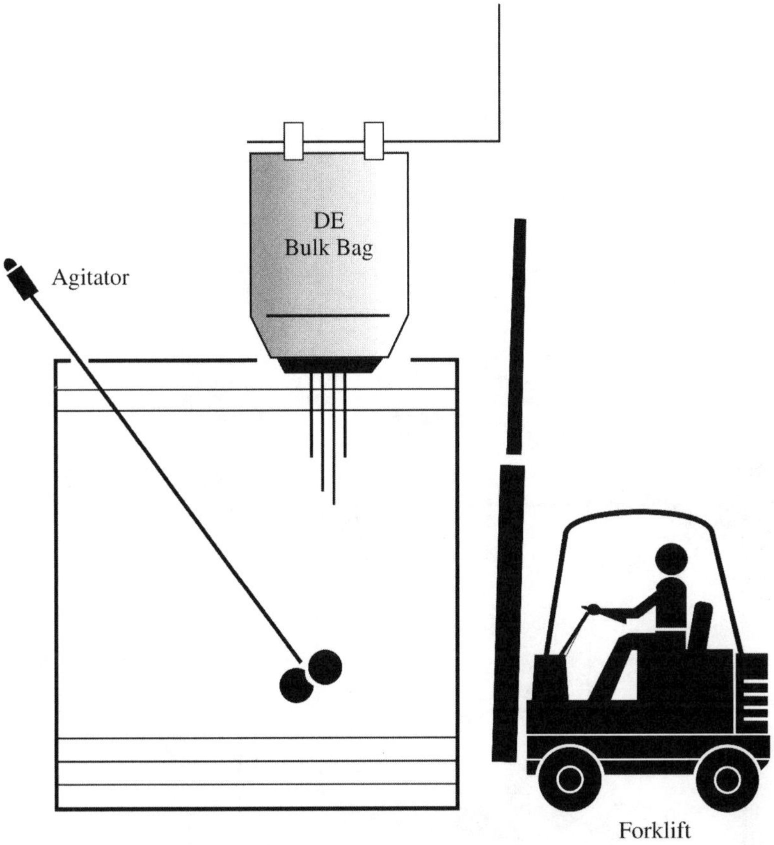

FIGURE 4-7. Gravity/Manual Discharge Direct to Slurry Tank

than 13 days. If the requirements are 1,000 lbs a day, more than five pallets would be required each week. For this use, the 900–1,000 lb bulk bag may be more practical. The benefits of each type of delivery would have to be evaluated for the use rates between these two limits.

Fifty-pound bags may be stacked two pallets high for a total height of close to 12 feet. Pallets are roughly 36 in. × 52 in. in size. Bulk bags will stack somewhat over 8 feet high on a 48 in. × 48 in. pallet. The type bag and method of storing (height) will determine the type of forklifts required. The bulk bags require more sophisticated unloading frames and forklifts, but significantly reduce the wear and tear on the operator in the manual lifting of 50-lb bags.

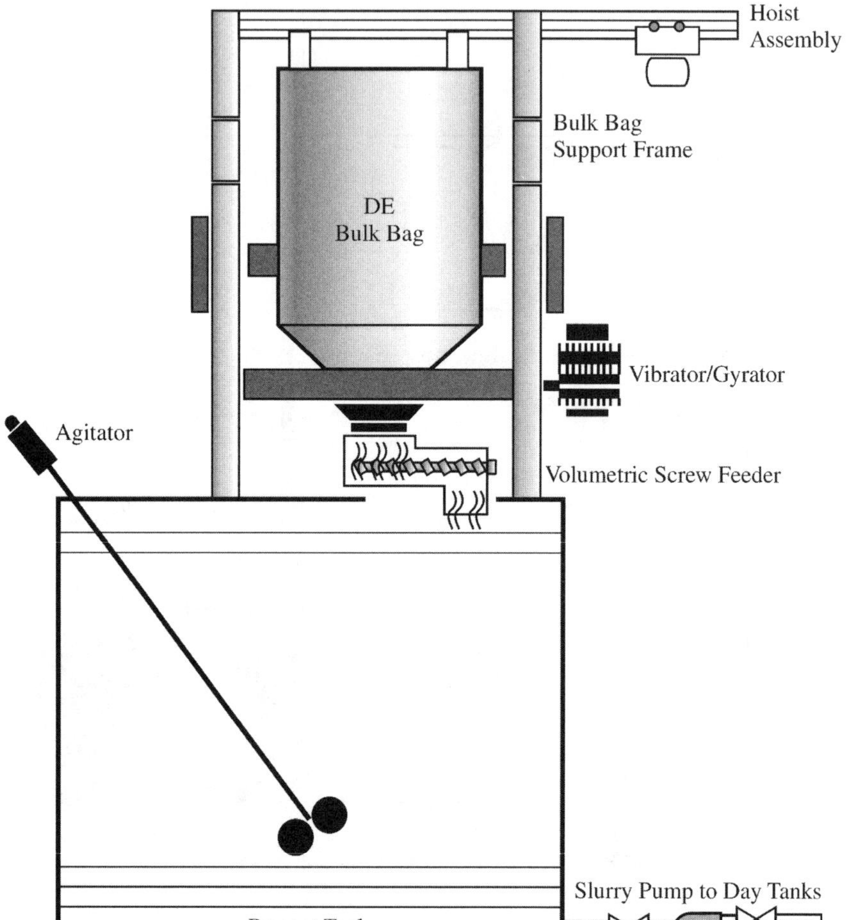

FIGURE 4-8. Bulk Bag—Vibrator/Gyrator-Assisted Unloading Direct to Slurry Tank

Diatomaceous Earth Slurry Storage and Transfer

Storage

DE slurry storage may be divided into two basic groups. DE in day tanks used for precoating and body feed may be maintained at 2 to 3% concentration to improve the rate of feed control. DE held in storage for distribution to several DE filter lines may be concentrated up to 20% by weight to reduce storage volume.

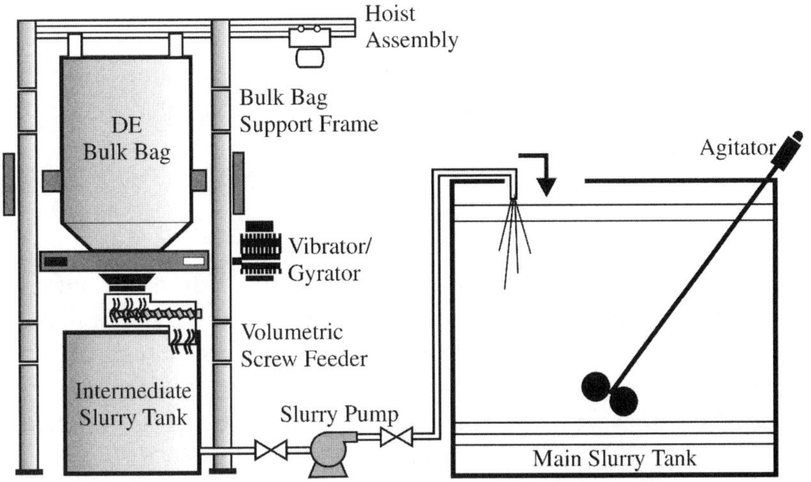

FIGURE 4-9. Bulk Bag—Vibrator/Gyrator-Assisted Unloading to Intermediate Slurry Conveying System

Once mixed in slurry form, diatomaceous earth must be maintained in suspension to inhibit settling. Tank mixers should provide sufficient circulation to lift solids off the bottom while generating a vertical velocity in excess of the settling rate of the largest DE particles. Too much mixing energy may cause DE particle wear and breakdown in size. The greatest wear occurs with higher concentrations of slurry. Tests have shown that there is barely noticeable wear and breakdown of DE after a week's time with concentrations up to 20%. In any event, it would be advantageous to limit storage of DE to three days.

Pipeline Conveyance

Research has indicated that there is more attrition of DE in suspension in pipeline conveyance than in storage mixing. As a rule of thumb, velocities to avoid unnecessary DE wear should be as follows:

— Concentrations up to 8%: less than 8 fps.
— Concentrations over 8%: less than 5.5 fps.

Velocities to avoid settling in pipelines should be above 2.5 fps. The precautions observed in the design of pipelines carrying suspended solids are also pertinent for diatomaceous earth slurry.

— Except for downward discharges at point of use, minimize vertical runs in pipe systems.
— Maintain minimum velocity to avoid particle settling. Settling rates of the larger-sized DE particles in the particular brand should control design (Table 3.4).
— If the pipeline flow is terminated after completion of transfer or interrupted by emergency, the pipeline must be flushed out immediately.

The aforementioned basic rules are particularly important for the higher concentrations of slurry (above 8%). A transfer line is most often flushed from the suction of the transfer pump to the end of run. For proper control of slurry concentrations, the diluted flushed material should be conveyed to a waste tank. Waste tanks should also be fitted with mixers to maintain slurry suspension. If virgin slurry is the wasted material, it may be accumulated and decanted for concentration for reuse at a later time. If the flushing water contains a minimum amount of DE, it may be used in the subsequent slurrying of dry material. It is always important to know the approximate particle concentration in reused waste flow to avoid errors in process.

One way to keep track of the DE concentration is to meter the inlet flow of clean water used for flushing. If most of the piping system used for conveyance can be sloped so that it may be emptied to the waste tank by gravity, the diluted DE concentration may be calculated from the volume of the piping system, the DE concentration originally conveyed, and the metered amount of flushing water.

Where DE may be recovered from slurried wastes in larger capacity plants (to be discussed later), the pipe-flushing water may simply be added to the wastes for eventual recovery of the DE.

Recirculation Distribution

The recirculation system is an ideal method for distributing DE slurry to multiple points of use. As shown in Figure 4-10, the three basic elements of a recirculation system are a storage tank with mixer for DE slurry, a recirculation pump, and a pipe loop serving all the points of use. If day tanks are to be filled for precoat and body feed at the points of use, the DE slurry may be in concentrations of up to 20% solids. Each day tank would be filled to the level desired by the operator and then diluted with filtered water to the 2 to 3% concentration usually used for precoat and

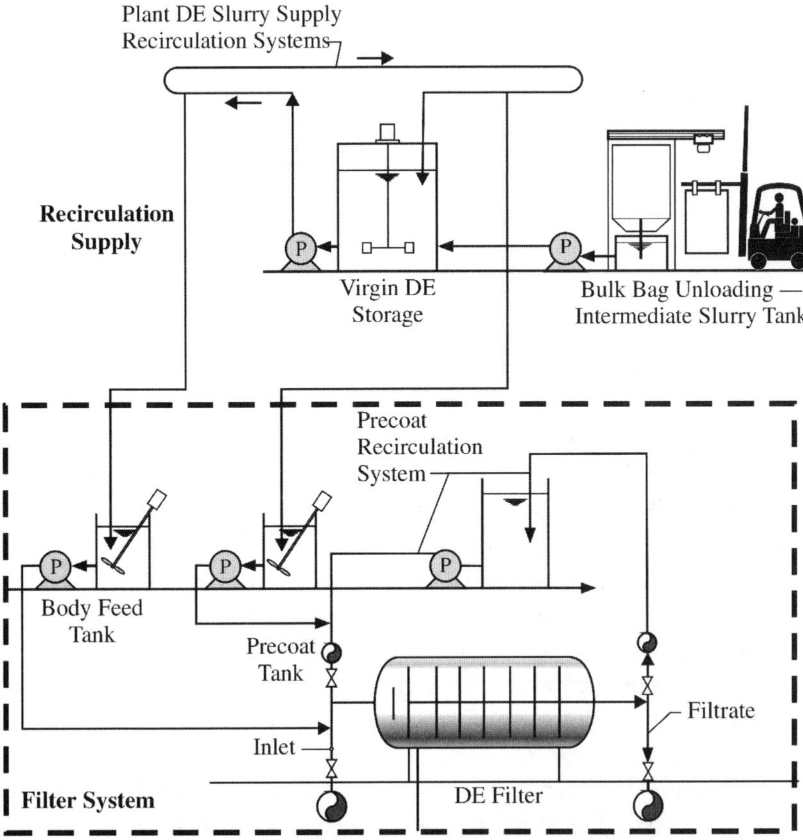

FIGURE 4-10. Diatomaceous Earth Slurry Recirculation Schematic

body feed. It is also possible to eliminate body feed and precoat day tanks by connecting the respective feed pump suctions directly to the recirculation line. In this alternative, a lower concentration of DE would be circulated that would be that required for the feed conditions. Multiple or spare recirculation systems would always be provided to ensure continuous service. As in the case of any DE slurry system, flushing provisions would be important to the operation of a recirculation system.

Distribution of concentrated DE slurry to day tanks would be an intermittent operation. Circulating the more dilute slurries for direct use would be a continuous operation. Single loops may be maintained in continuous operation, but multiple storage tanks would be required to avoid interruption of continuity of supply. As a rule, once flow is interrupted in a circulating line, the line should be flushed before returned to service. This flushing may not be necessary for simple pipe systems (minimum pipe joints and upward vertical runs) using dilute slurry. Flushing is always necessary following interrupted flow of concentrated flow. To conserve the use of water, it is advisable to use the slurry recirculation pump during flushing by feeding the water into the pump suction.

Diatomaceous Earth Recovery and Recycle

A particular advantage of diatomaceous earth filtration is that the DE may be recovered from the sluiced slurry wastes for recycle as body feed in larger capacity plants. It is emphasized that, for drinking water, recovered DE would only be used for body feed, not for precoat purposes. Virgin DE would always be used for precoat, which is the final buffer in the filter cake before discharge of the filtrate.

The wastes collected from the end-of-run flushing of the filter consists of the source water solids removed in filtration, DE body feed, and the DE precoat material. Separation of the lighter source water solids fraction from the more dense diatomaceous earth volume is accomplished by pumping the wastes through a series of vertically mounted batteries of hydrocyclones.

Because of the erosive nature of the diatomaceous earth in the slurry, it is necessary to use hydrocyclones of ceramic construction and pumps of abrasion-resistant aerial. The ceramic hydrocyclones used are most often the 1-inch size with an upper limit of 2 inches. Larger-sized units would be impractical because of the cost of making larger customized ceramic units. The size limitation of the hydrocyclones necessitates assembly of individual units in batteries connected to pipe manifolds.

Waste slurry is pumped under pressure into the battery manifolds to enter the top of each hydrocyclone in a horizontal direction, tangential to the entry chamber. The swirling centrifugal action that is generated will force the heavier diatomaceous earth particles against the inside perimeter of the hydrocyclone, which is tapered to a smaller diameter at the bottom discharge. The heavier diatomaceous earth solids will gravitate to the bottom discharge, which empties into a collector tank. The source water solids and the finer fraction of DE particles, lighter in weight, are left in the center core of the hydrocyclones. This lighter fraction is forced out of the top of the hydrocyclone by means of the pressure induced by the inlet rate of flow and the tapered bottom discharge. These waste tailings are collected for further hydrocycloning or are discharged to waste.

Because of the necessity to remove the maximum amount of pathogens that may be present in the source water solids, the hydrocyclone recovery of DE is accomplished in several successive steps. Recovery systems are usually purchased completely assembled, mounted on skids or platforms, fabricated by a single manufacturer. Design details of recovery systems are, therefore, beyond the scope of this text. An example of a typical recovery system is shown in Figure 4-11. As indicated, there are four successive steps of DE recovery in which the DE concentration increases as more water is removed.

A scavenger step may be added to this particular arrangement to improve overall efficiency by recovering additional DE from the tailings of the somewhat inefficient first hydrocycloning step. As shown on the schematic, the lower volume tailings from the second through fourth steps are passed back for reprocessing to the previous step. In this way, the only waste stream for the system would be the tailings from the first stage or, where added, the scavenger hydrocyloning.

As indicated, dilution water may be added to the collection tank in the third step, which feeds the last, fourth set of hydrocyclones. This water is needed if the concentration of DE becomes too high for efficient operation of the final hydrocyclones.

Somewhat more than 90% of the DE in waste discharge from the filters may be recovered. With some loss of the finer fraction of DE particles, the recovered material is slightly coarser than the original product. Research has shown that the recovered material can be more effective for

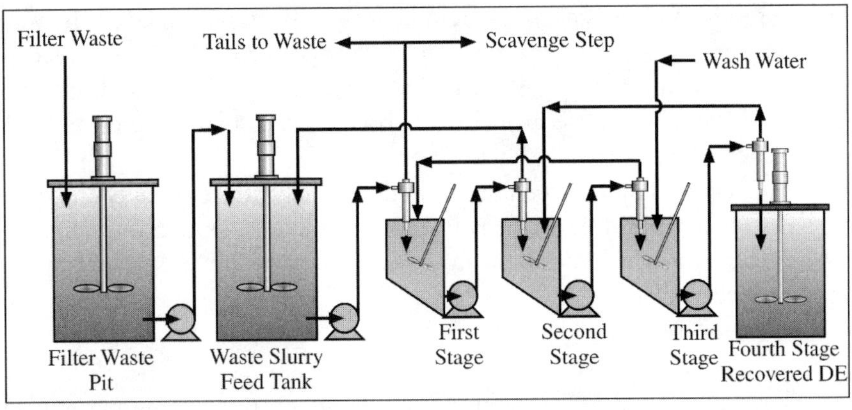

FIGURE 4-11. Diatomaceous Earth Recovery System Schematic

body feed than virgin materials. Actually, hydrocycloning improves the uniformity coeffient. With the loss of the finer fraction, the porosity of the cake is improved, but not to the extent that breakthrough is a concern. On the average, length of filter run with recovered material has been at least 10 to 15% longer.

With a loss of more than 15% of the DE, makeup with virgin material may be necessary. Some, if not all, of the makeup may be provided by the virgin precoat at the start of each run. Depending on the nature of raw water particlate matter, recovery of DE may significantly reduce the cost of operating a DE filtration plant. Where silt-like particulates predominate in the raw water, carryover with the recovered DE may inhibit re-use. The use of DE recovery would also be limited by the capacity of a particular DE filter plant and the amount of body feed required. Careful evaluation must be made of the cost of installing and operating a DE recovery system compared to the alternative of greater costs for the purchase of DE and ultimate disposal of wastes. With DE recovery, there is a parallel reduction in the volume of wastes for disposal. Evaluations may show effective cost reduction for recovery systems that would be operated intermittently a few times a week.

Challenge studies have shown that DE recovery systems are effective in removing the *Giardia* and *Cryptosporidum* cysts. Nevertheless, the larger sizes of these pathogens of concern may pass through with the recycled material so that disinfection before re-use is always necessary. Fortunately, recovered DE can be kept in storage for several days' time with a nominal increase in plant costs for the increased storage. It is known that with a high enough concentration of potassium permanganate, *Giardia* cysts can be inactivated in 24 hours. *Cryptosporidum* inactivation time is not known at present, but would be longer. If ozone is used in the water treatment train, cyst inactivation in the recovered DE may be accomplished with a relatively short period of ozonation. This final disinfection would also inactivate any virus or bacteria that may be present in the recovered material.

Test runs must be conducted to determine both the quality and quantity of recovered DE that may be expected before adoption for design.

Waste Disposal

The wastes from DE filtration are a mixture of the source water solids in addition to the DE precoat and body feed that is sluiced from the filter leaves. Source water solids will amount to less than 5.0 mg/l, most often less than 2.0 mg/l. With body feed added at rates of from 10.0 to 50 mg/l, the proportion of DE including the precoat is many times that of the source water solids. While the DE particles are usually much more diverse than the source water particulate matter, the size of the DE particles make it difficult for separation and waste treatment by simple settling.

Table 3.4 in the previous chapter shows the particle size distribution in Hyflo media, typically used in drinking water treatment, along with the calculated respective settling velocities in water. For effective separation in a settling basin, the overflow or upflow rate at the top must be less than the settling rate of the particles to be removed. For instance, if all DE particles 5 microns and larger were removed in a clarifier, the overflow rate would have to be less than 0.000032 ft/sec (Table 3.4) or 0.014 gpm/sq. ft. This would be an extremely low overflow rate to use for design of a waste clarifier.

The settling velocities shown in Table 3.4 are for discrete particles. In a homogeneous mix, the larger particles actually would tend to drag down some of the smaller-sized DE. Referring to the Table, about 50% of the Hyflo DE particles range between 8 and 24 microns in size. Settling rates for these sizes range between 0.000082 and 0.000789 ft/sec or 0.036 and 0.035 gpm/sq. ft. With less than 25% of the particles less than 8 microns in size, it would be reasonable to assume that between 80 and 90% of the solids might be separated in a clarifier designed with overflow rates of between 0.10 and 0.20 gpm/sq. ft. In fact, it has been demonstrated that sluiced slurry of Hyflo media left in a basin overnight will settle sufficiently to permit significant decanting of supernatant. The most advantageous method of separating solids from the sluiced slurry flow, however, would be to use plate settlers or lagoons (where area is available). In general, plate settlers designed for a projected overflow rate of less than 0.10 gpm/sq. ft. would provide sufficient first-stage separation. Since plate settler designs will vary by manufacturer, it would be best to submit the gradians of DE particle sizes (similar to Table 3.4) for a particular waste along with the waste flow discharge rate and solids concentration to the different manufacturers to determine the plate settler design required.

Figure 4-12 shows a block diagram of the equipment and flow paths for treating typical DE filter sluiced slurry wastes. All waste flow would first pass through a plate settler fitted with rotating bottom collector/thickener mechanisms. The thickener paddles help free the water bound in the solids, which rises to the overflow. The overflow will contain

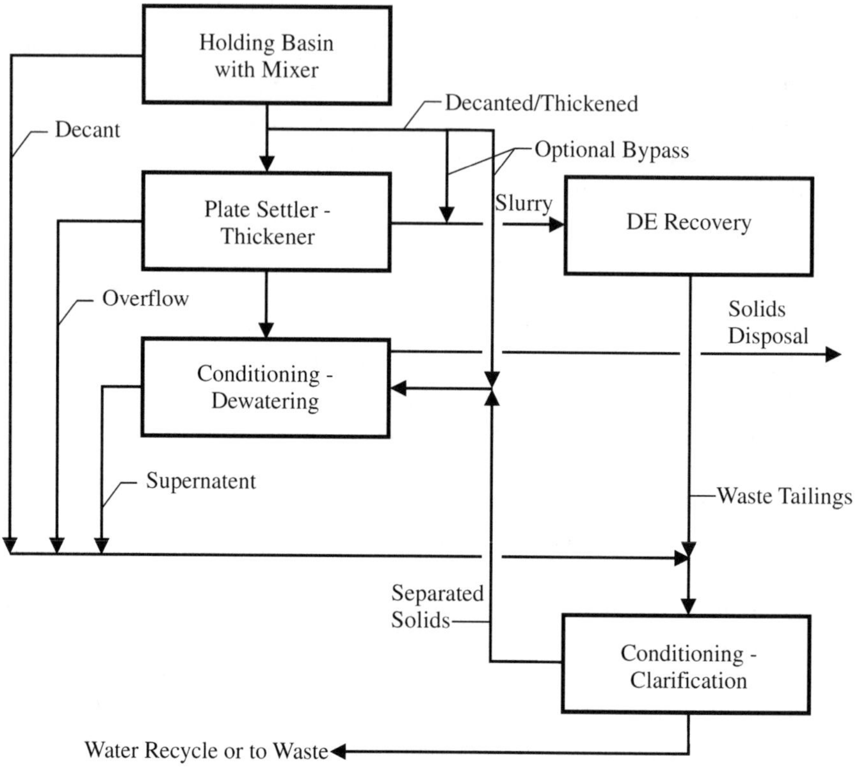

FIGURE 4-12. Diatomaceous Earth Disposal Options

the finer DE particles along with source water solids that pass up through the plates.

As indicated in the Figure, the concentrated DE solids may be used in recovery or may be conditioned and dewatered for ultimate disposal. All decanted flow along with waste tailings from the plate settler thickener, DE recovery, and dewatering would be collected for coagulation and clarification. Solids from this final clarifier would be cycled to dewatering. The overflow may, if permitted, be recycled to the head of the filter plant or may be disposed to waste. It is noted that this last clarification step may be eliminated if disposal of the tailings to the local sewerage system is permitted.

Concentrated DE may be dewatered using conventional vacuum filters or belt presses. The design and specification details of the equipment required for treating sluiced slurry wastes are beyond the scope of this text. It may be stated, however, that more conservative design factors would be used for plate settler separation, but that the separated material dewaters much more readily than those from conventional rapid sand filter plant wastes.

While functionally practical, the use of lagoons and drying beds for DE wastes should be considered with caution. Once dried, the powder-like DE can be easily disturbed and lifted by light winds.

5
Pilot Testing

Despite decades of efforts, it has not been found possible to accurately predict filter performance or to develop specific filter design criteria simply from evaluation of chemical/physical source water data. Moreover, there are few situations in which a new application of DE filtration may have to be provided that çan be considered to be identical to other operating diatomaceous earth filter facilities. In drinking water treatment it is not unusual to find two filter installations taking supply from the same body of water at different locations that exhibit profoundly different treatment requirements.

Some bench and/or pilot scale tests should be performed before proceeding with diatomaceous earth filter process design. Such tests are small-scale simulations of full-scale systems. The basic tests should generate sufficient data to provide answers to the following questions:

— Will DE filtration produce a potable water of acceptable quality?
— Is DE filtration cost effective when compared to other methods of treatment?

Simulation

The key to effective pilot tests is simulation. The scale of the tests or the size of equipment used is not as significant if the filtering elements are performing in the same manner as the prototype or full-scale unit. The following two DE filter test arrangements for diatomaceous earth filters have proven to be effective in providing the answers to one or both of the questions posed previously:

— *Buchner Funnel.* Using this common laboratory filter equipped with graded Whatman paper along with other conventional laboratory fittings and accommodations, meaningful tests can be conducted on a bench scale. The driving force for filtration flow is the vacuum applied to the underside of the filter paper.

— *Square Foot Filter.* This single-leaf unit has a total of one square foot of filter area and has appurtenances that can accurately parallel full-scale pressure filter operation or a small-scale unit to test vacuum filter performance.

Several pilot test assemblies are available from DE filter manufacturers or producers of DE media. These are usually small-scale versions of the manufacturers' product line. Some of these units are as small as one square foot filters; others, like those for vacuum filtration, may simply be the smallest units manufactured, having individual filter leaf areas of up to 10 square feet. Some care must be exercised when using some of these proprietary filters. Often the arrangement of the filters or the mode of operation can influence how test data may be applied to full-scale design. Operating data from a vacuum type test unit, for instance, would not be particularly useful for the design of a pressure DE filter system. The pilot test arrangement should simulate the anticipated full-scale system as much as possible.

Principal Process Issues

Filters are designed to remove particulate matter. Two principal mechanisms, entrapment and attachment, are involved in the separation of suspended solids carried in the flow stream. In DE filtration, the major portion of particulate removal is accomplished by entrapment. Separation by entrapment is a straining process in which the particles are caught in the interstices or openings between the filter media grains. In this action, most of the particles are separated at the surface of the accumulating cake as the flow enters the filter media.

Particles small enough to pass through the media openings are separated by attachment. The suspended particles are attracted and held to the media grains by means of chemical-physical forces. Included in such action are the Van der Waals forces in which the relative attraction is influenced by the charge characteristics of the separated solids and media particulates.

As suspended matter is separated and stored on or in the media during a filter run, the separated particulates become, in effect, part of the media. Over time, particulates accumulate and gradually influence filter performance. Such influence on filter performance depends upon certain particle characteristics as follows:

— Small-sized particles that can pass through media interstices and tend to depend more on attachment for separation.
— A broad range of particle sizes that tend to fill up the media openings more completely; the smaller sizes fit into the pores of media and between other particles to cause rapid head buildup and reduce length of runs.
— Compressible particulates like algae or other similar organic matter, where the increasing force applied to the filter by the hydraulic head differential builds up and compresses the collected matter, tending to close up the pores or interstices to reduce the length of run.

Where media blinding or significant loss of porosity is the principal problem, increasing the rate of body feed helps to maintain media porosity. As diatomaceous earth body feed of the same or somewhat coarser grade than the precoat is fed to the incoming flow to the filter, a matrix is provided in the solids buildup that literally "holds apart" the raw water particulates to inhibit blinding.

Media conditioning and pretreatment of the incoming water have proven to be effective in a number of instances in the improvement of both the removal of colloidal-size suspended matter and the reduction of color. In media conditioning, alum is mixed in the precoat and body feed DE slurries. Alum coating of the media provides the positive charge necessary to neutralize and then attract negatively charged colloidal particles.

While ozone is not a flocculent in the true sense, ozone can initiate coagulation or coalescing in certain situations. Ozone added to the incoming water will both destroy complexes linking organic matter and react with iron or manganese. In reaction, the metallic ions can be oxidized to form oxide and hydroxide precipitates. Although these precipitates may tend to be somewhat fragile, tests have shown that they can be readily removed by diatomaceous earth filtration. The ozone-precipitation process may, therefore, be considered to be somewhat analogous to coagulation where source water solids are coalesced to enhance removal.

Process Variables

Several process variables should be evaluated in the development of both a pilot test program and the resulting prototype design criteria. These variables may be considered to be both adjustable and fixed. Process system features like the type of septa cloth or the grade of diatomaceous earth may be changed in pilot tests. The grade of DE or the amount used for precoat or body feed may also be changed in full-scale operation for changing process conditions. It would not be practical, however, to change the filter cloth in full-scale operation. Most often this filter feature remains fixed after initial adjustments.

The most significant process feature over which there is little control is source water quality, especially if such source is a surface water. Surface waters are subjected to seasonal changes, influenced predominantly by temperature and runoff fluctuations. These constantly changing influences will impact on the type and quantity of particulate matter suspended in the flow withdrawn for treatment. It is important in a test program to determine how the controllable features of treatment may be adjusted without causing upset. The major process variables are discussed later by category.

Surface Water Quality

The principal influences on changes in surface water quality are rainfall/runoff, temperature, and limnological activity in reservoirs and lakes. Rainfall/runoff influences the amount of sediment detritus and mineral matter that may be washed down into the streams and carried in the flow. The increased flow may also cause the upset of previously settled sediment. Temperature changes may create environments particularly advantageous for the growth and profileration of various species of algae, plants, and other biological organisms. Temperature will also affect operation of the filter as it affects changing water viscosity.

The impact of surges in natural runoff and stream flow on changes in source water particulate content is obvious. Unstable ground surface soil and detritus as well-settled matter in streams and lakes can be disturbed and carried away if the hydraulic forces are sufficiently turbulent. Both suspended and dissolved matter in source waters are thereby increased. The impact of temperature is more subtle. A rising temperature may cause settled biological matter to decay and putrify. This action may eventually produce gases which in release will cause particulate and dissolved matter to be introduced into the water above. Dissolved matter could have deleterious color, taste, and odor characteristics. Dissolved iron and manganese can also be released by such temperature-caused bottom action.

The limnological activity in reservoirs and lakes generate changes in the types and concentration of organic matter. In the development of water quality standards, organics are becoming more important. Organic content in surface waters may be divided into particulate, colloidal, and dissolved fractions. The particulate and colloidal fractions are significant to DE filter operation.

The initial origin of organics that enter water supply systems is the decay of plant and tree detritus on watersheds. With rainfall-runoff activity, products of decay wash out of the detritus or leach out of the soils and wetlands and are either transported overland to streams and/or percolate through the overburden to enter the groundwater flow. When this flow reaches lakes and reservoirs, the detention provides a dynamic environment in which the biotic communities further affect both the magnitude and nature of the organic content of the water.

The characteristics of organic matter in drinking water sources are as different as the sources available. The variations in quantity and quality features are greatest in stream and river flow. Such variations tend to be moderated in lakes and reservoirs because of the detention and mixing provided, as well as the biological activity. To say that the nature of limnological and other processes influencing the generation of organics is extremely complex would be almost an understatement. While these processes and the features of natural organic matter (NOM) may be discussed in general terms, it is emphasized that each individual source water has unique organic characteristics that demand sufficient research to determine the drinking water treatment required for all anticipated variations.

Figure 5-1 shows the variation in turbidity and color for the City of New York Croton water supply for a single year. The incidence of bottom upset due to rising water temperatures is obvious in the peaking of color and turbidity values in the fall. Similar peaks have also occurred at other times in other years due to unusual rainfall and runoff conditions. It is noted that for this supply, color changes closely parallel changes in total organic content (TOC), which is not shown.

Different species of algae require different combinations of water temperature, sunlight, and organic nutrients for growth and proliferation. Nutrients may be contributed by overland washdown of fertilizers or compost materials. Nutrients may also be recycled by upset of bottom sludge. Natural water flow variations through a body of water will control the amount of organic nutrients recycled and the degree of dilution.

Sunlight will impact both surface water temperature and plant growth itself. The sunlight needs of different plant species will vary. The combination of water temperature, sunlight, nutrient content, and the

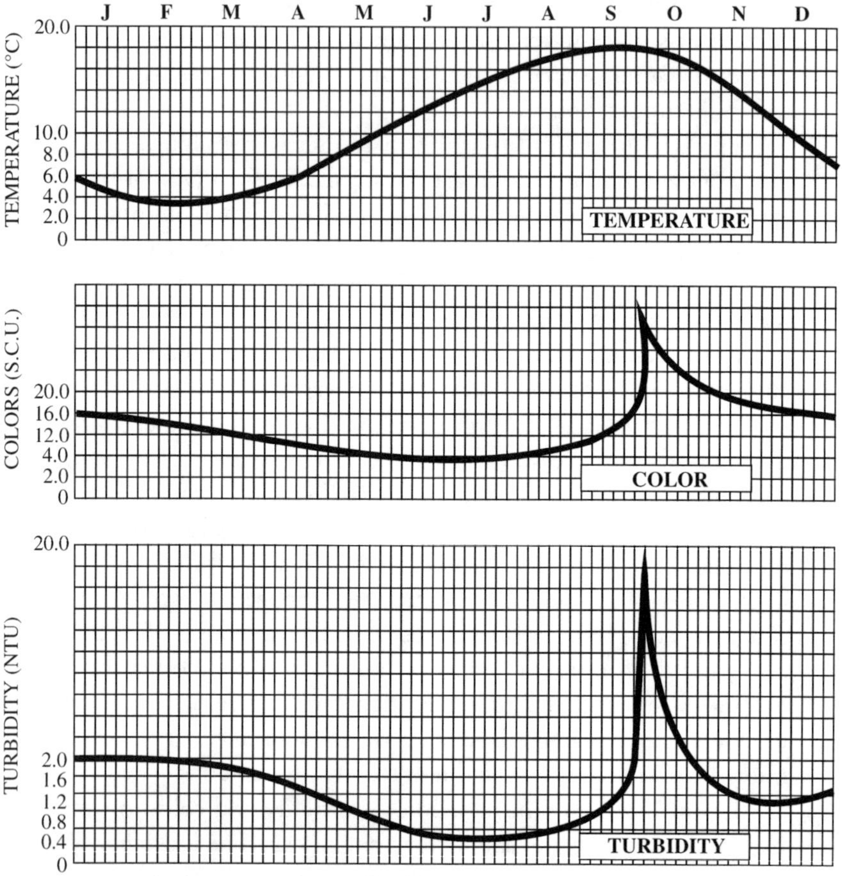

FIGURE 5-1. Source Water Quality—Seasonal Variations

season of the year will influence both the type and the amount of plant or biological growth that may occur in a body of water. Insect larvae and other water organisms can also proliferate under similar influences.

As cold temperatures approach freezing, the oxygen saturation potential of water is significantly increased. In vacuum type DE filters, the pressure on the downstream or exit side at the cake may be reduced enough to cause the release of dissolved oxygen or air. Such air release or air "popping" can disturb the cake and cause DE to enter the filtrate and raise turbidity.

It is obvious that the source water variables can have profound impact on operation of diatomaceous earth filter operations. For meaningful design of filter facilities it is necessary to know which type or which season of raw water quality may limit filter capacity most, or what feature will require special pretreatment conditions. It would also be desirable to determine how variations in raw water quality may affect operation. It may be possible, for instance, to utilize only part of the filter capacity provided for seasons of the year where the source water solids content is greatly reduced or is of a type that is more readily removed by the DE filter. During these periods, the filters may be operated at higher application rates.

Groundwater Sources

DE filters would be used in the treatment of groundwater or well supplies predominantly for two purposes.

1. To remove pathogens where short-circuiting from an adjacent surface stream or body of water has been demonstrated
2. To remove iron and/or manganese with the addition of preoxidation.

The need for body feed will be almost negligible where the water is being filtered just for pathogen removal. Here, the use of finer DE media might be considered to further improve removal.

The treatment variations for iron and manganese removal would be influenced mostly by the content of these contaminants and the surface area of suspended DE particulates required to promote caking and adsorption.

Filter and Media Relationships

The septa cloth and the DE media are the process variables that must be evaluated in system design. The septa are made of cloth woven with metallic or plastic strands. Strand diameter, pattern of weave and effective openings are the septa variables where some choice is available. The filter equipment manufacturers have had considerable background experiences in a multitude of applications. By relying on this experience, it would be possible to limit or reduce tests for a specific application to a few selections of septa cloth.

The more economical precoat materials are the finer grades of diatomaceous earth. Coarser grades have higher production costs. Head loss is greater through the finer precoat, hence pumping costs to develop higher hydraulic heads would be greater. Any test program should start with the finer grades and work up through the coarser material until an advantageous balance of DE and pumping costs is reached.

The separated solids must also be considered to be part of the functioning filter. Once the precoat is covered, filtration occurs on the surface of the solids accumulated in the filter cake. In the development of a filter design for a specific application, the following should be the principal considerations:

- — Septa should have openings sufficiently small to hold the bridged precoat layer through the filter run without occurrences of particulate breakthrough for the most efficient grade of media.
- — Cake, made up of precoat and separated solids along with body feed, will provide paths of flow small enough to entrap sufficient suspended solids to produce the required quality product water in terms of residual turbidity or residual particle count.

The addition of body feed is effective in reducing the potential for filter blinding and shortened lengths of runs. Body feed intermixed in the cake with the accumulated raw water solids forms a matrix to increase the porosity of the overall filter cake. The variables in body feed selection are the grade and the amount of DE used. Naturally, as greater amounts and coarser grades of DE are used, the higher the material costs. This must be balanced against the benefits of extended filter runs and lower pumping costs.

Pretreatment

The needs and variables for pretreatment prior to DE filtration would be source water specific. Because the use of DE filters is somewhat limited to treatment of water with low-level suspended solids content, the types of pretreatment that may be employed is somewhat limited. The principal pretreatment utilized in DE filtration is oxidation of NOM to reduce and eliminate the natural negative charges of colloidal particulates to enhance removal in the filter cake. This may be accomplished by means of aeration or through the application of ozone and chemical oxidants like potassium permanganate. Oxidation is also affective in the precipitation of dissolved matter like iron or manganese or for reducing the content of organic constituents like color, tastes, and odors, or possibly the trihalomethane formation potential.

Pilot Program Design

The primary purpose of pilot research is to first determine the functional and economical viability of diatomaceous earth filtration compared to alternative treatment. Then it would be important to determine whether DE filtration is a viable treatment for compliance with existing as well as anticipated drinking water regulations. Certainly, establishing criteria for ultimate design and operation would be basic to any research program. Development of such criteria might also be a requirement of a state regulatory agency for project approval.

While the pilot program may be a relatively inexpensive portion of a treatment improvement project, careful planning is essential to avoid loss of time and unnecessarily increased costs. For instance, unusual seasonal water quality treatment problems may occur during pilot plant operation for which insufficient equipment and materials have been provided to perform or conduct meaningful process tests. This can mean that a year will be lost until the same seasonal quality features occur again. Such delay not only will delay implementation of a treatment improvement program, but also could be costly because of inflationary construction cost trends. It is therefore essential that the design of the pilot program include flexibility to allow simple and easy modification of the test facilities and procedures. Every effort must be made in the review of background raw water quality data to aid in the anticipation of potential treatment problems or treatment impacts. Some of the more important features of pilot program design are discussed later.

Purpose

The particular needs for a pilot test program will vary significantly from project to project. A summary of such purposes, including those mentioned before, might be as follows:

- Determine cost and treatment viability to permit meaningful comparisons with other treatment processes.
- Provide information for regulatory review in sufficient detail to satisfy the regulatory authorities and obtain process approval.
- Develop design criteria in sufficient detail to ensure a cost-effective final design in terms of both construction and operating charges.
- Simulate potential problems like the infrequent occurrence of giardia cysts. This may require "spiking" the raw water with surrogate cysts to simulate such potential problems.

— Develop supplemental treatment needs like the use of ozone or other oxidants to improve color reduction or the reduction of the THM formation potential.

Drinking water quality requirements are continuously being developed by the federal EPA in accordance with the Safe Drinking Water Act. As much as possible, potential changes or additions to the water quality regulations should be anticipated in a test program design. There may well be, for instance, a more stringent requirement for THM content below the present 0.08 mg/l maximum level. Even if THM levels comply with present requirements, it would prove helpful for future use to investigate the possibility of reducing the THM formation potential in a pilot test train. Tests might be conducted using BAC for pretreatment with or without oxidants. Disinfectants other than chlorine could also be investigated. This information could be obtained at relatively low cost in an operating pilot train. Otherwise, the pilot train would have to be re-installed or reactivated at much higher cost in the future when the change in regulation takes place.

Data Needs

Sufficient data should be generated in the pilot program to satisfy the essential objectives of the investigation. Much more data than actually required may be developed so nothing important is overlooked. This generation of possibly redundant data will occur mostly in the laboratory analysis of test samples. Obtaining too much test data not only can be costly, it also can complicate test evaluation where it becomes difficult to "see the trees for the forest." Conversely, generating too little data may necessitate repeating tests at increased costs.

Development of a list of data needs starts before the design of the program with the acquisition, collation, and review of available water quality background data. These data should include sufficient sample analyses to indicate present problems as well as trends in water quality. A source body of water may be in an advanced state of eutrophication so that water 20 or 30 years hence may have a greater THM formation potential or increased color levels. Evaluation of these data may also indicate the need to spike the raw water being tested to simulate possible future conditions. At the least, the evaluation of existing raw water data will indicate chronic existing quality problems, potential infrequent quality problems, and future trends that should establish both treatment needs and test data needs.

In essence, development of the principal goals of the pilot investigation will help determine the scope of the basic data needs. Potential changes in raw water quality and/or regulatory product water quality requirements will indicate where extension or expansion of the test program would be beneficial to provide data for future process accommodation.

Facilities

The basic diatomaceous earth filtration equipment arrangements that would provide the essence of full-scale system simulation are discussed later. These would include the Buchner Funnel bench test arrangement and the "square foot" or other filter system for pressure and vacuum tests. These systems would include all the elements necessary to test the filtration process. In many treatment situations, it will be necessary to provide pre-treatment or post-treatment to attain the required quality product or filtrate. Pre-treatment methods involve conditioning of the raw water flow or conditioning of the DE precoat media. Most of these involve the feed of oxidants like ozone or potassium permanganate to condition the inlet flow or the use of coagulants like alum to coat filter media. Except for gaseous ozone, the pre-treatment materials would be fed to the incoming water flow or to the body feed supplement by means of proportional metering pumps. Gaseous ozone would be added through some type of diffuser system to permit sufficient contact time of the ozone bubbles with the water flow. Ozonation and some of the chemical or slurry additions may, therefore, require a contact tank that would provide sufficient detention time for the desired action or reaction to take place. It is noted that the use of pre-treatment like oxidation may require the addition of BAC columns before filtration to reduce BDOC and the THM formation potential.

Descriptions of the various methods of post-treatment are also beyond the scope of this handbook. In general, these will vary from the alternative methods of disinfection to stabilization of the product water to inhibit corrosion.

Some general accommodations must be made available for all DE pilot systems such as a drainage system to accept the continuous flow of product water and some means for disposing of the waste cake discharged in cleaning. The size and other features of the space available to install the pilot system would also be important. Evaluation of the purpose of the study, the preeminent features of treatment, and the data needs of the investigation will indicate what laboratory analysis conveniences should be provided at the test site and what analyses could be conducted by an off-site laboratory. Unfortunately, the facility selection and design will, in many instances, be influenced by available budget. Where funds are limited, special care must be exercised in determining which features of the pilot installation are most important.

Other Program Features

The remaining features of program design include the following:

- — Operating and laboratory staff
- — Test alternatives and schedules
- — Data collection and analysis
- — Project documentation
- — Interpretation of findings.

The first three items cited previously are somewhat closely interrelated. The numbers of alternative tests planned, the seasonal variations in raw water quality that must be handled, and the data collection and analyses that evolve will determine staff requirements. Such requirements will include both the number of staff and their qualifications. Type of test equipment employed and the possibility of extended filter runs will determine whether the test facility must be operated "around the clock," five or seven days a week, or whether it can be on a "9-to-5" weekday schedule. Each program will have its own unique needs.

Project documentation is a most important aspect of the overall program. Such documentation would include those features of the test that are not just pure data. For instance, a sudden change in raw water quality might cause a seriously shortened filter run or, possibly, filter breakthrough. The raw water data might show enough to indicate cause and effect. On the other hand, it would also be important to check out the system when such treatment breakdown might occur. It is possible that a body feed system could have failed due to lack of supply slurry or due to feed pump clogging. Such system scrutiny and information would be an important part of the overall test reporting. Many times, the observation of the operator during changes in filter performance will provide more significant background clues for evaluating the tests than the numbers results.

Interpretation of the findings is probably the most critical feature of the pilot program. The limitations of the various data or information gathered must be carefully considered. Some analysis data may be developed with less precise methods than others. A "one time happenstance" should not be permitted to influence major decisions unless the background information is irrefutable. This is the stage of the study in which "red herrings" can surface to steer the results in the wrong direction.

It is important that trends be carefully evaluated. Such trends might include the change in filter performance (effluent quality and length of run) that results from a gradual or rapid change in raw water quality. Graphical plots of the trends that occur with change of grade of DE used for precoat and body feed or the change in the rate of body feed additions might also provide important evidence on which to develop prototype design criteria. In general, the interpretation of findings involves a careful evaluation of cause and effect that includes all the data and information developed in the tests.

Buchner Funnel Test

The Buchner funnel is a laboratory device that is traditionally used to determine the filterability of slurries or sludges. It is a glass funnel fitted with a wire screen support on which Whatman or similar type filter papers in various grades are used for diatomaceous earth filter tests.

Vacuum, readily available in most laboratories, is used to provide the driving force to generate flow through the Buchner funnel. The funnel may be fitted into stoppers that seal ordinary glass laboratory flasks or cylinders. Vacuum is then applied to the underside of the funnel through a connection through the same stopper.

The Whatman paper simulates the filter septum. The glass equipment enhances observations of the filtration procedures. While the Buchner funnel elements and operations do simulate diatomaceous filtration, the data developed cannot be the primary basis to use for full-scale design.

The two Buchner funnel tests that are most often conducted serve principally as indicators of the qualitative and quantitative effectiveness of diatomaceous earth filtration. These two tests provide a quick, inexpensive means to determine 1) whether diatomaceous earth filtration will produce the clarity of filtrate desired and 2) at what approximate flow rates and with what rates of body feed the filtration process might be operated. In addition to determining the effectiveness of DE filtration, the Buchner funnel tests can also be used for preliminary determination of the grade of diatomaceous earth to use in more accurate pilot filter simulations discussed later. The bench test can indicate the approximate grade or grades of DE that should be used in the pilot arrangement in much shorter time than it would take to go through several pilot runs to arrive at the same answers.

General Test Features and Rules

The Buchner funnel test is conducted in batches, each individual batch in the order of 1.0 liter volume. Test procedures are somewhat different from actual filter operation in that the grades of diatomaceous earth being tested are mixed directly into the batches of liquid suspension to be filtered. In actual filter operation, precoat is accomplished first, before the flow of liquid suspension is introduced. In the Buchner funnel tests, the precoat and filtration tests are almost a simultaneous operation since the DE is mixed into the liquid suspension being tested. Some initial filtration is accomplished as the precoat is being formed on the Whatman paper.

While these tests are not exact filter simulations, some basic filter rules must be followed. Most important is the rate of liquid that must be manually poured into the funnel during the batch run. A liquid level must be maintained over the septum or cake at all times. It essential that the integrity of the formed cake not be disturbed either by the turbulence caused by the introduction of flow or by permitting too low a liquid level that will allow air to be sucked through the cake. The cake will crack if permitted to become dry.

Qualitative Tests

In the qualitative tests, diatomaceous earth is mixed into the liquid batches to be tested in approximately the same proportions used in filter operation. Rate of flow application is not important in these tests other than to preserve the integrity of the cake. To produce the required quality, it may prove beneficial to increase base amount of DE added. These tests determine the grade, not the amount of DE that will produce the desired quality. For the quantitative tests, the DE additions would more closely simulate anticipated body feed additions.

The results of these tests will answer two basic questions:

1. Will DE filtration produce the desired clarity of filtrate?
2. How fine a grade of DE must be used to produce such clarity?

Figure 5-2 is a schematic arrangement of equipment for the qualitative Buchner funnel test. Filtrate flask A receives the filtrate until the precoat is completed. Flask B receives the final or product filtrate flow. As shown, a No. 4 Whatman paper would be used as the filter septum. Vacuum would be applied to either flask by means of the valves or tube clamps provided. In the arrangement shown, the Buchner funnel is connected by a "tee" fitting to permit directing flow to either of the two flanks. Most often, flexible rubber or plastic tubing is used for interconnections.

As shown, a throttling valve is used to regulate the vacuum applied, which is indicated by the gauge. Usually, this vacuum is maintained somewhat higher than 20 inches of mercury. This vacuum is possible only after the precoat is established that is necessary to provide resistance to flow. Several trial runs may be necessary before the desired vacuum setting is ascertained. The vacuum should not be so high that the high flow generated will disturb the cake. It should be recognized that determination of the DE

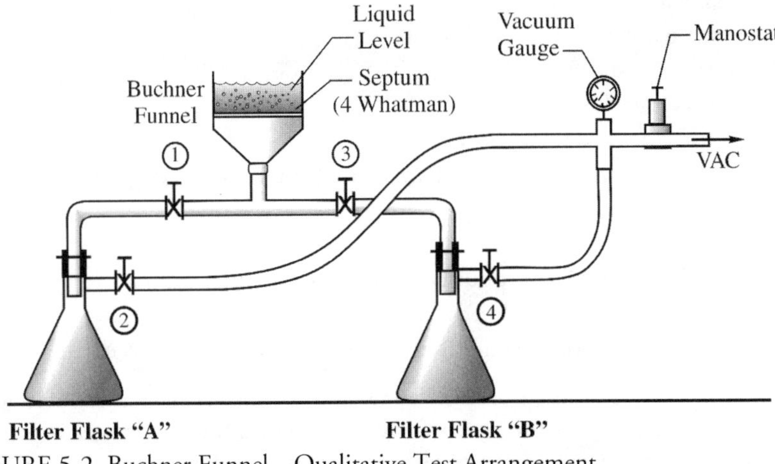

FIGURE 5-2. Buchner Funnel—Qualitative Test Arrangement

grade to be used is a primary part of this test. Difficulty in maintaining vacuum can be a factor of DE grade. Too coarse a grade may make it difficult to hold desired vacuum at high enough levels. Too fine DE grades may result in an extremely low rate of filtration at the highest vacuum available. The intent of the tests is to determine the coarsest grade of diatomaceous earth that will produce the desired clarity or quality of filtrate.

Table 5.1 shows the sequence of procedures for the qualitative tests. In summary, flow is directed to Flask A until a precoat is formed when flow is switched to Flask B. This switch is the critical part of the test. If too much time is taken during the switch, it is possible for the liquid level to be drawn down to the cake level. The time permitted to make this switch can be increased by starting with a high level in the funnel and/or by reducing the vacuum to reduce the rate of flow-through.

Data developed would be grade of DE vs. clarity (turbidity) or filtrate quality achieved. It is important that the coarsest grade DE that can produce the required quality be applied in the subsequent quantitative tests described later.

TABLE 5.1. Buchner Funnel Test, Qualitative: Clarity vs. Diatomaceous Earth Grade

Procedure

1. Stir admix diatomaceous earth (DE) into liquid to be filtered (about 4 to 8 grams per liter for 4-in. diameter funnel).
2. Close all valves.
3. Turn on vacuum.
4. Open valves 1 and 2.
5. Pour slurry into funnel.
6. Adjust vacuum to desired flow.
7. When enough slurry is filtered to form a precoat on the septum, open valves 3 and 4.
8. Close valves 1 and 2.
9. Remove filter flask A and pour contents into Buchner Funnel.
10. Analyze filtrate in flask B for clarity.
11. If necessary, repeat test using different grade of diatomaceous earth.

Notes: The liquid level in the Buchner Funnel must be kept as high as possible at all times to avoid disturbing the precoat as new slurry is added. If the filter cake is allowed to become dry, cracking will occur and filtrate clarity is not valid.

Quantitative Tests

The primary purposes of the quantitative tests are to determine the following:

— Need for DE body feed
— Grade of DE body feed
— Rate of DE body feed.

Table 5.2 lists the general procedures for these tests using the Buchner funnel arrangement shown in Figure 5-3. Starting with the grade of DE found beneficial in the qualitative tests, tests would be conducted using different grades of diatomaceous earth at different feed or mix rates. Again, the DE is mixed in with the liquid or water sample to be tested.

TABLE 5.2. Buchner Funnel Test, Quantitative: Flow Rate vs. Body Feed

Test Procedure

1. Stir admix diatomaceous earth into liquid to be filtered (15 to 100+ mg/l depending on nature of water solids).
2. Close valve and adjust vacuum to desired level. Pour slurry into funnel.
3. Open vacuum valve and start timer.
4. Record time and volume filtered at regular increments for DE grade and amount added.

Analysis and Calculations

5. Plot time vs. cumulative volume on log/log graph paper (time as the Abscissa).
6. Extrapolate to desired cycle length.
7. Record volume in ml at extrapolated time (V1).
8. Convert final volume (ml) into gal/sq. ft. (V2).

$$V_2 = \frac{V_1 \times (144/A)}{3{,}785}$$

Where:

V_1 = final volume in ml
V_2 = gal/sq. ft.
A = area of funnel in square inches.

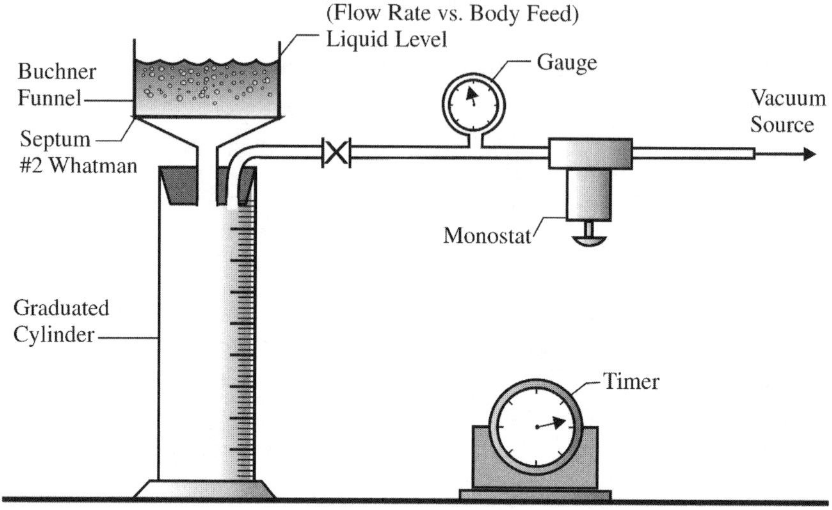

FIGURE 5-3Buchner Funnel uantitative Test Arrangement

The total volume of the graduated cylinder is used to record the incremental amounts of filtrate accumulated over time.

For these tests, both the grade and the proportion of DE used would be varied to determine that combination (DE grade and proportion) producing the desired rate of flow application to the filter. In the analysis and calculation steps, the plot of volume vs. time would be projected to the desired cycle length. For water treatment, this would be 12 hours or more. The volume filtered (V1) would be determined from this projection and converted to gallons per square foot of septum area (V2) using the equation shown in Table 5.2. To determine the rate in gpm/sq. ft., V2 must be divided by the cycle time selected (in minutes) on the extrapolated plot.

The differential pressure used in the quantitative test is a vacuum, usually around 20 inches mercury (20″ Hg). As a rule of thumb, 2 inches of mercury vacuum (2″ Hg) equals a differential pressure of about one pound per square inch (1.0 psi). If the volume or rate of flow desired is at a higher differential pressure, the adjustment would be made using the following:

$$V3 = V2 \times \frac{(pf)^{1/2}}{(pt)^{1/2}}$$

Where:

V2 = test flow (gal/sq. ft.)
V3 = adjusted flow (gal./sq. ft.)
pf = final desired differential pressure (in psi)
pt = test pressure differential (in psi).

Again, the flow volume per square foot of septum may be converted to gpm/sq. ft. by dividing the incremental time (in minutes) from the extrapolated plot.

The Buchner funnel tests do not develop data sufficiently accurate for design. At best, these tests provide enough information to determine whether the DE filter would be a viable consideration for the particular source water application. For the most part, both qualitative and quantitative (production) performances of the full scale DE filter will be superior to that determined in the bench tests. This bench test does, however, provide valuable information at a low cost in a short time to enable decisions on proceeding with the more extensive pilot tests later.

Fundamental Pilot Filter Tests

Because of the number of variables that may exist in a pilot test situation, the filter tests must proceed in several steps or phases. The intent would be to study or determine one feature of the filter operation in each step. The fundamental tests or steps in any pilot simulation program would be to determine the following in the order shown:

1. Grade of DE for precoat
2. Optimum amount of DE for body feed
3. Cost-effective filter application rate.

Information obtained in each test would be used in the subsequent steps. In other words, in the second step body feed determinations, the tests would employ the grade of precoat established as the coarsest to yield the quality of filtrate desired. The third series or filter application rate tests would employ both the precoat and body feed data established before. As will be discussed later, it may prove to be advantageous to refine some of the information developed in the three stages cited previously in a final series of tests. Another filter variable that exists is the septum cloth. For the fundamental test series, the septum cloth used in the pilot filter would be that grade or type in most common use for filtration of similar-quality water. After the grade of precoat is determined it may prove to be advisable to use either a closer or a more open weave of cloth.

Precoat

The primary intent of these first series tests would be to determine the grade of DE that will maintain the highest rate of flow while producing acceptable clarity or product water turbidity. The approximate grade of precoat DE may first be established using the Buchner funnel test. For situations in which such test data or other background experience is not available, the first test should be conducted with a DE grade like the Hyflo Super-Cel. Successive tests would move up and down the size range as indicated by the results. If sufficient clarity is not attained, finer grades of DE would be employed in successive steps until acceptable filtrate clarity is produced. If acceptable clarity is accomplished in the first test, coarser grades would be used until breakthrough occurs. It is noted that in the precoating cycle, the filtrate should clear up in from 2 to 5 minutes' time. Since this does not necessarily indicate that the precoat is all in place, the precoat recycle should continue until the liquid flow observed in the recycle tank is clear. This usually takes place in 5 to 10 minutes at most.

For these first test series, it is recommended that a precoat of 20 lbs/100 sq. ft. (0.20 lb/sq. ft.) be applied and a body feed (same grade DE) be admixed at a rate or ratio of 5:1. The filter rate would be at 1 gpm/sq. ft. of filter. The body feed (in mg/l) may be calculated as the ratio of body feed addition to source water solids (as NTU turbidity). The test cycle for potable water should be at least four hours, with clarity checks and notations at 5-, 15-, and then at 30-minute intervals. Acceptable clarity (<0.5 NTU) should be attained within this time after the precoat is established. As mentioned, the DE precoat material selected would be the coarsest grade producing acceptable clarity.

It is cautioned that breakthrough or insufficient clarity in a particular test may be caused by improper or incomplete precoat or by the quality or characteristics of the water being tested. The reason for turbidity breakthrough, discussed at the end of this chapter, should be established before proceeding with the next test with a finer grade of DE. If certain source water characteristics, like the presence of colloidal or organic material, limit the success of DE filtration, other treatment measures may have to be added. This is discussed later.

Body Feed

The intent of the second series tests is to determine the optimum body feed amount or rate of addition to the inlet water flow. For potable water filtration, optimum body feed may be defined as that amount (or rate) that produces the maximum yield expressed in gallons of filtrate per pound of DE body feed. Expressed another way, optimum body feed is that rate of addition producing the longest run (in time) for the available increase in head loss.

For these tests, the DE selected in the first series is used for both precoat and body feed. Precoat should be applied at the same 0.20 lbs/sq. ft. of septum area. The filter rate would be maintained at the same constant 1 gpm/sq. ft. as before.

In this series of tests, total throughput vs. the amount of body feed rate must be determined. This means that flow is maintained until a maximum pressure differential is attained. For pressure type DE units in water filtration application, this maximum pressure differential may range up to 35 lbs/sq. in. In many specific situations, this differential pressure may be established by what may be available from a reservoir source by gravity in relation to water system pressure. For lack of such definitive data, the most common maximum pressure differential of 25 lbs/sq. in. may be used for the tests. Data would be tabulated as total throughput of flow as gallons or cubic meters for the rate of DE body feed used as mg/l. Dividing the throughput by rate of body feed will show gallons produced per unit of body feed. The higher the number, the more efficient the use of body feed. It would also be helpful to tabulate the pressure drop through the filter at fixed increments of time through the filter run. This plot of pressure drop vs. time often provides a useful picture of filter performance.

To find that most efficient rate of body feed, tests may be initiated using 10 mg/l of body feed to yield a rate of production per unit of DE used. In the next test, the rate of body feed would be increased, say to 15 mg/l. If the yield per unit rate were to be less than the previous test, the subsequent test might employ 7.5 mg/l of DE body feed. The body feed adjustment in each test should track the optimum rate in this manner. Body feed should be increased or decreased in the direction in which the production rate appears to increase. Figure 5-4 shows the plot of total throughput vs.

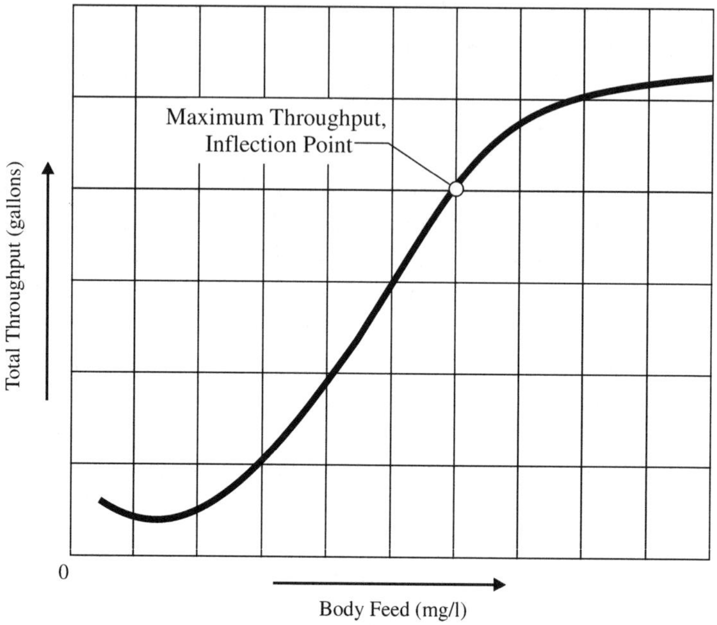

FIGURE 5-4. Total Flow Throughput vs. Body Feed

body feed rate over a broad range. Maximum throughput for this series of tests is at the inflection point of the rising curve as shown on the figure.

Filtration Rate

Up to this point, the intent has been to determine feasible and efficient usage of diatomaceous earth. The filter rate selected has been a somewhat arbitrary, although a typical 1.0 gpm/sq. ft. Since it is the filtration rate that determines the size and costs of the filter installation required, it would be advantageous to run a third series of tests on filter rates. Results of such tests would be used to estimate relative equipment, costs, and operating costs to determine the most feasible compromise between the two. These estimates would necessarily be site-specific. That is to say, local labor, power, and material costs would have to be used for the evaluations to be meaningful. In locations where power, costs are low and shipping charges for DE high, the economic influences would be different than where power costs might be high and shipping charges less.

In any case, the routine to be followed for the filtration rate tests would be the same. The optimum values of DE grade and precoat from the previous two test series would be used and only the filtration rate would be varied. Pressure drop vs. time data would be recorded and plotted for analysis.

As a guideline for these tests, the theoretical change in cycle length should first be calculated using the relationships discussed in the unit process chapter. If a filter cycle is 40 hours at an application rate of 1.0 gpm/sq. ft., the cycle would be reduced to 10 hours' length at the increased rate of 2.0 gpm/sq. ft. Expressed another way, if it required 40 hours to filter 200 gpm of water through a 200 sq. ft. filter at a rate of 1.0 gpm/sq. ft., use of a 100-sq.-ft. filter at the same 200 gpm inlet rate would shorten the filter cycle to 10 hours. These examples serve only as guidelines; data obtained from the filtration rate series will determine the actual relationship of filter runs vs. rate. Operating experience has indicated that precoat filter performance will exceed that predicted by the theoretical relationships to produce longer runs at increased flow application rates.

The range of pressure filter application rates that may be used for drinking water treatment may be generally limited to from 0.5 to 4.0 gpm/sq. ft. Vacuum filter rates of more than 1.5 gpm/sq. ft. are rarely practical because of the limited head loss. As mentioned, the 1.0 gpm/sq. ft. rate has been used in past installations. It is suggested that the initial filtration rate test be operated at the 1.0 gpm/sq. ft. rate similar to the first two series of tests. Subsequent tests may be conducted with 0.5 gpm rate adjustments, say at 0.5, 1.5, and 2.0 gpm/sq. ft. filter rates to attain the desired length of filter runs. For large-capacity pressure filter operations, application rates of up to 4.0 gpm/sq. ft. may be considered. Even if the most cost effective or functionally desirable operating rate is obvious, it would be advantageous to complete the runs to determine the performance factors over the complete range. Water treatment systems most often are operated at changing rates of production to meet consumer needs. Emergencies can occur when some treatment units may not be available because of emergency repair. It would be helpful, therefore, to know what lengths of runs are possible using a filter at its nominal design rate of 1.0 gpm/sq. ft. and at a possible emergency rate of up to 4.0 gpm/sq. ft.

Supplementary Treatment

Sometimes it is necessary to provide additional treatment in conjunction with diatomaceous earth filtration to 1) improve solids removal performance or to 2) handle special source water problems like soluble iron and manganese, color, tastes, and odors. It is likely that improving solids removal may be associated with special or unusual source water quality problems. Some examples of additional treatment that may be employed as pretreatment or that may supplement the filter process itself are as follows:

— *Colloidal Solids Removal.* When colloidal solids are bound up with natural matter, treatment with ozone sometimes will destroy the bond and effect neutralization of the negative charge to permit improved DE filter solids removal. Ozonation may also reduce compressible solids.
— *Color Removal.* Color may be soluble iron or color may be iron or lignin bound up with natural organics. Ozone will precipitate soluble iron. Ozone will also destroy the organic bond and then precipitate bound iron or other matter for easier filtration removal.
— *Soluble Iron Removal.* Iron may also be precipitated from solution by mixing magnesite (MgO) with the raw water along with the body feed in a detention tank before the filter. The resulting particulates may then be readily removed by the DE filter.
— *Soluble Manganese.* Potassium permanganate ($KMnO_4$) will oxidize and precipitate manganese. It can be mixed in the raw water with provisions for some detention (usually 5 to 10 minutes) before DE filtration.
— *Taste and Odor Reduction.* Some tastes and odors may be reduced with the use of strong oxidants like ozone and potassium permanganate discussed before.

A detailed description of tests for the supplementary processes is beyond the province of this handbook. Scrutiny of these processes, however, shows common requirements. Thorough mixing of chemicals or additional substances used with the raw water is essential to ensure as complete a reaction or contact as possible. Adequate detention time is also a necessary requirement for most of these supplementary processes. Mixing and detention may be provided in the water treatment system pipelines, in additional detention tanks, or in the DE filter enclosure itself.

Square Foot Pressure Filter

The square foot filter provides close, if not exact, simulation of prototype pressure diatomaceous earth filter operation. That is to say, individual equipment items are assembled into a system that matches full-scale designs. The square foot filter system includes individual pumps for raw water feed and body feed addition. For the most part, flow rate will range from 0.5 to 4.0 gpm for the one square foot filter area. This permits measuring the flow continuously during the filter run. Significant features of the major components that would make up a square foot filter system are shown in the Figure 5-5 photograph and in the Figure 5-7 schematic. The more important aspects of such components and operation are reviewed later.

FIGURE 5-5. Square Foot Filter

Filter

The filter consists of a single filter leaf having a total septum area of 1.0 sq. ft. For the unit shown, the leaf is installed in a cylindrical transparent shell, held between two stainless steel flanges. Connections are provided for inlet, outlet, drain, and interior sprays for backwashing. Piping connections are 1/2-in. stainless steel. While the filter can withstand a working pressure of 60 psi, the maximum pressure for drinking water tests would normally be lower. A pressure drop limit across the filter cake of 25 to 35 psi is the practical limit for most drinking water diatomaceous earth filtration applications. Some pressure drop occurs through a "clean" pilot system before the filter cycle starts with precoating. It is noted that other types of square foot filters can be obtained from various manufacturers. Most of these consist of a rectangular leaf installed in a vertical stainless steel or acrylic cylindrical shell with a removable head.

Filter Feed Pump

The ideal filter feed pump is one that will deliver a constant flow rate with no pulsations to disturb the filter cake. Usually, centrifugal pumps are used for full-sized diatomaceous earth filter systems. The square foot filter requires a pump that will deliver flow rates adjustable between 0.5 gpm and 4.0 gpm. Pumps most suitable for such pilot application are the rotary pump with flexible impellers or the tube (peristaltic) pump. Both are positive displacement type pumps, but their advancing cavities are relatively small in size so the pulsations produced are slight. Both the rotary and tube pumps should be provided with variable speed drives to produce the rotating speed required for the desired flow-rate.

As pressure across the filter increases with buildup in cake thickness, some decrease in pump flow rate will occur. This decrease is, however, slight and may be corrected by periodically opening the filter inlet valve to compensate for the increase in pressure across the cake. The drop in flow rate with these type pumps is usually in the order of 10% with an increase in discharge pressure to 35 psi from an initial 10 psi. The filter inlet valve should be somewhat throttled at the start of a filter run to permit reducing the pressure drop across the valve later to compensate for filter cake loss. It would also be possible to operate with no such adjustment if the drop in flow rate from start to end of run is 10% or less. It will be necessary to calibrate valve settings and pump rotation (speed) before the tests are initiated to ensure proper flow conditions during the pilot runs.

Body Feed Pumps

It is also helpful to use a body feed pump that minimizes a pulsating flow. Because of the low flow rate required, the peristaltic or plunger diaphragm metering type or pump would be the most desirable. As shown in Figure 5-6, the body feed rate may be adjusted by varying the DE concentration in the feed slurry or by varying the rate of feed or flow. The plots in Figure 5-6 are for a filter inlet feed rate of 1.0 gpm. For lower or higher filter application rates, the body feed mg/l must be adjusted proportionately. Body feed is a separate flow added to the filter inlet as it is in full-sized sys-

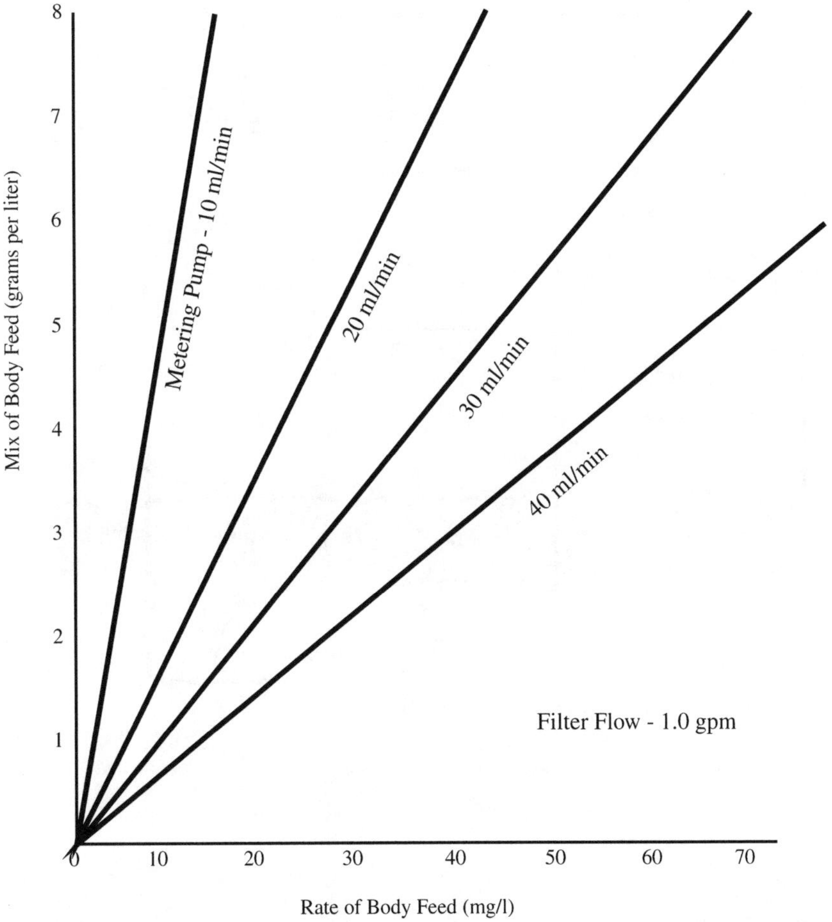

FIGURE 5-6. Body Feed Rate vs. Dilution Options

tems. Since no flow meter is provided for this rather low flow rate, the pump rate must be checked by measuring the drop in level in the body feed tank over a period of time. Flow rate calibrations and adjustments should be made during an actual filtering cycle.

Instrumentation and Controls

As shown on the schematic diagrams for the precoat, filtering, and cleaning cycles (Figures 5-7, 5-8, and 5-9), the following instrumentation must be provided.

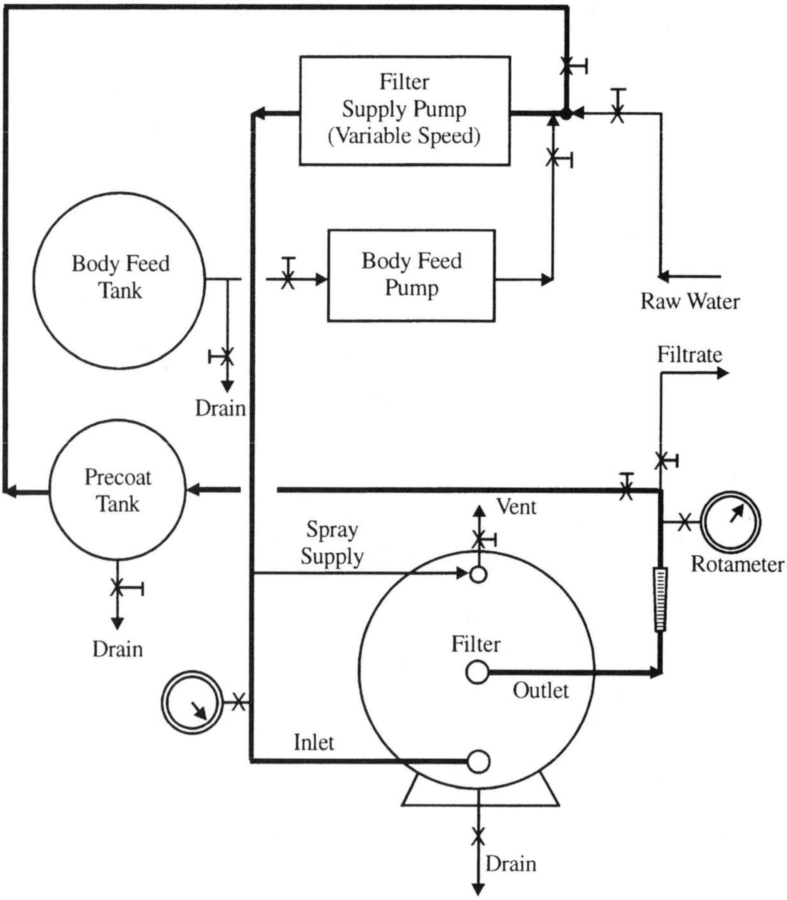

FIGURE 5-7. Pilot Filter—Precoat Flow Cycle Schematic

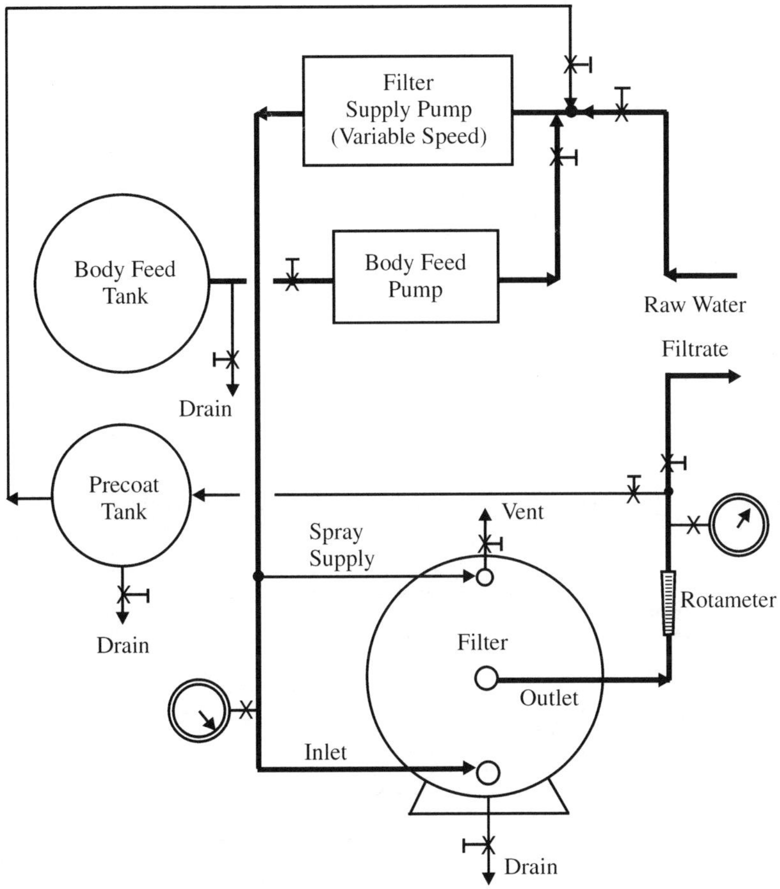

FIGURE 5-8. Pilot Filter—Filter Flow Cycle Schematic

— Flow Meter: usually a rotameter with a flow range of from 0 to 10 gpm
— Filter Pressure Gauge: 0 to 50 psi range
— Spray Cleaner Gauge: 0 to 50 psi range.

The use of the filter inlet valve to compensate for increased loss of head across the accumulating cake was discussed before. The spray cleaner inlet valve is also used as a throttling device to control such flow. In this case, the valve is used to maintain the spray inlet pressure to about 40 psi. A turbidimeter is furnished as part of the square foot filter system to continuously monitor filtrate turbidity. Depending on the meter characteristics, either the full flow or a side stream from the effluent filter line will

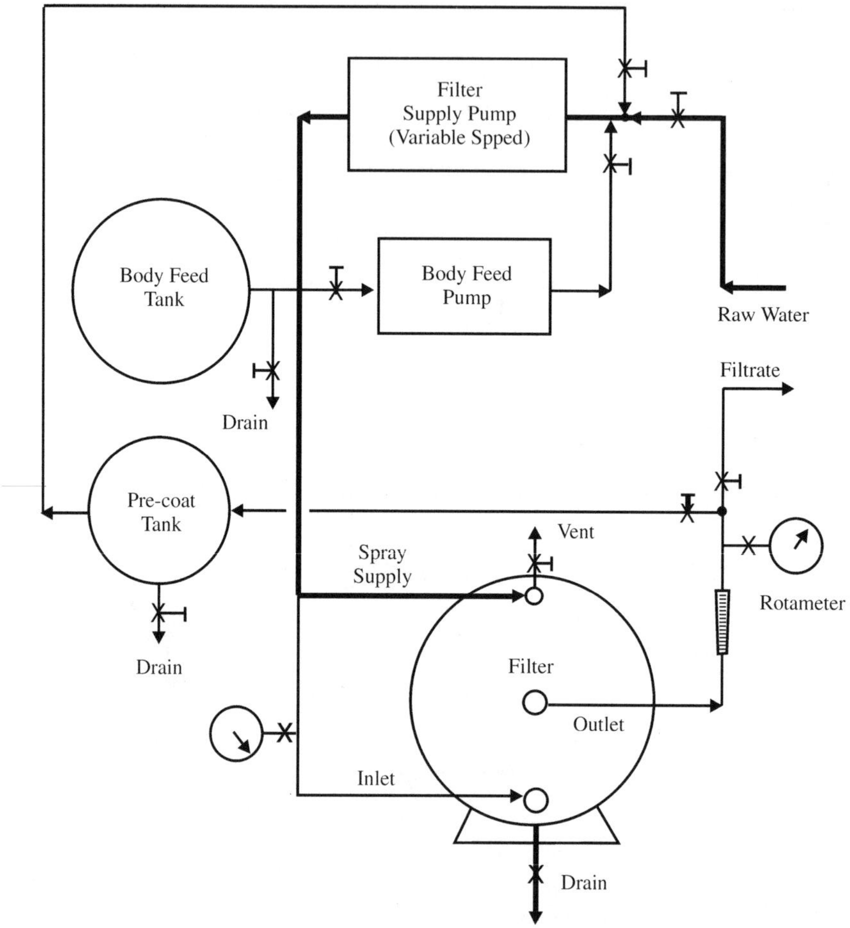

FIGURE 5-9.Pilot Filter—Filter Cleaning Cycle Schematic

pass through the unit. A particle counter for continuous or bench top determinations would also be desireable to determine the effectiveness of relative particle size removal.

Tanks and Tubing

At least two tanks must be furnished to hold the precoat and body feed slurries. The precoat tank connected in the recirculating piping should have its top surface above the filter. Body feed volume is determined by the concentration of the slurry and the length of run anticipated. In general, the capacity of the two slurry tanks will vary between 5 and 15 gallons. Tanks may be of polypropylene or stainless steel construction. Continuous operation electric powered mixers must be provided for each tank to maintain the DE in a relatively homogenous suspension.

Tubing may be used to interconnect the equipment in the test system. Tubing may be polyethylene, vinyl, or other good quality flexible plastic. Tubing used for tube pumps should be as specified by the pump manufacturer. Fittings should be of rigid polypropylene or nylon construction. Pump discharge lines should utilize compression-type fittings or tight clamps joints designed to withstand the higher pressures. Valves should be the ball type of PVC or polypropylene construction.

Test Procedures

Filter operating procedures are shown in Tables 5.3 through 5.6. The filters would be operated in the same manner for each test run. Differences between test runs would exist for:

— Grade of DE used
— Amount of precoat in lbs/sq. ft. of septum
— Rate of body feed.

TABLE 5.3. Square Foot Filter, Pilot Test: Preliminary Setup

1. Close all valves in the system. There are 12 valves in all (see Schematic Figure 5-7).
2. Make sure all hoses are connected. Hosing and connection points are shown on the schematic.
3. Set all switches on the filter unit to the off position. Set the pressure controller on the control box at 30 psi* and place a new chart on the recorder. Turn on the master switch and the pump control switch.
4. Thread the Tygon tubing through the body feed pump and turn the wing nuts all the way on the pressure plate. The size tubing is selected based on flow rate.
5. Fill the body feed tank with 50 liters water and fill the precoat tank to the black line (use drinking water or filtrate from a previous run).

*Or other maximum pressure setting (20-40 psi).

TABLE 5.4. Square Foot Filter, Pilot Test: Precoat

Figure 5-7 shows the flow circuit for precoating. Be sure all other valves are closed.

1. Weigh out filter aid to be used for precoat and body feed. The amount of filter aid to be used for body feed can be taken from Figure 5-6 based on the desired flow rate of the body feed pump and the dosage in the influent.
2. Open the following valves (see Figure 5-7):
   - Precoat inlet valve
   - Filter inlet valve
   - Precoat recycle valve
   - Air vent.
3. Turn on the feed pump.
4. When the filter shell is full, close the air vent and set the pump flow rate.
5. Add the precoat filter aid to the water recirculating in the precoat tank. Using a paddle or large spatula, agitate the slurry continuously for about five minutes, then occasionally until the water in the shell is clear.
6. While waiting for the precoat to go on, add the filter aid for body feed to the body feed tank and start the mixer. Open the body feed tank valve. Disconnect body feed inlet tubing; direct discharge to graduated cylinder.
7. Start the body feed pump. Adjust the speed to the flow rate selected by use of a graduated cylinder and stopwatch.
8. Turn the body feed pump off and reconnect the tubing to the body feed inlet valve.

Note: Once the precoat cycle has started, you cannot shut down the feed pump. If the pump should shut down during precoat or filtration cycles, the filter must go through the cleaning cycle and be precoated again.

TABLE 5.5. Square Foot Filter, Pilot Test: Filtration

The flow pattern for the filtration cycles is shown in Figure 5-8.

1. Slowly turn the speed adjuster on the feed pump down, adjusting the flow to desired flow rate (usually start at 1 gpm and change later if desired).
2. Open the body feed inlet valve and start the body feed pump. Run the body feed for about two to three minutes.
3. After the body feed has run for two to three minutes (still in the precoat cycle), check to be sure that all valves are open on the influent line up to but not including the influent valve.
4. Open the influent valve and close the precoat inlet valve. (ALWAYS OPEN ONE VALVE BEFORE CLOSING THE OTHER VALVE.)
5. Open the filtrate valve and close the precoat recycle valve. Trace the flow pattern. It should match up to Figure 5-8.
6. The filter cycle will continue until 1) the operator shuts down or 2) the pressure regulator shuts the system down at 30 psi. The pressure regulator is also designed to shut down the pumps if there is a power outage.

Note: As stated in the Precoat Section, if the pumps shut down for any reason, you must clean the filter and precoat again before going to the filtrate step.

TABLE 5.6. Square Foot Filter, Pilot Test: Cleaning

The flow pattern for cleaning is shown in Figure 5-9.

1. Open the body feed drain valve. Keep the stirrer on so that the DE will not settle to the bottom of the tank.
2. Open the filter drain valve and air vent (in that order).
3. Fill the precoat tank with clean water (drinking water or filtrate).
4. Open the precoat inlet valve; close the influent valve and the body feed inlet valve.
5. Start the feed pump. Turn the speed adjuster up four turns (there should be a fair amount of agitation indicated on the bottom of the filter shell).
6. Open the spray cleaner valve and the spray cleaner gauge valve. Set the pressure regulator about 45 psi.

In summary, the sequence of tests should be as follows to determine the optimum performance conditions:

1. Grade of precoat DE
2. Rate of body feed addition
3. Filtration rate.

Pilot Vacuum Filter

In addition to pumping arrangement, the basic difference between the pilot pressure filter and a pilot vacuum unit is size. It is possible to achieve close simulation of full-scale operation with a square foot pressure filter or even a unit with as small as a half square foot leaf. Vacuum pilot units, on the other hand, are just low-capacity vacuum filters in contrast to the almost miniaturized pressure pilot simulation.

Most often pilot vacuum filters will have one or two leaves, each ranging from 6 to 10 square feet in area. This means that the rate of flow in a pilot vacuum test might range from 5 gpm to 20 gpm. Almost any type pump may be used to deliver water under pressure to the pilot pressure filter. The vacuum filter requires a pump that will develop the suction required for filter operations. The selection of small capacity pumps that would provide reasonable suction simulation for pilot vacuum filter tests is limited. The higher the rate of test flow, the better the selection and resulting simulations. This is another reason for operating vacuum filter tests at higher rates of flow.

Another option available for vacuum filter tests is the method of adding DE for precoat and body feed. While these additions may be made using metering pumps to feed a DE slurry suspension, a dry feeder may also be used. A simple volumetric-type feeder may be mounted over the inlet chamber to add the measured amount of dry DE for both precoat and body feed.

While the pump arrangement is different, the three cycles of precoating, filtering, and cleaning would be the same as outlined for the pressure filter. The basic differences in the operation of the pilot units would be the same as that inherent in full-scaled operation.

— The vacuum filter is open for observation to permit inspection of the precoat as well as the accumulation of cake in the filter cycle.

— The filter leaf or leaves would be cleaned manually using a high-pressure hose.

These two features provide a particular advantage for the vacuum filter tests. With the full operation open to view, responding to problems and general trouble shooting becomes more simple.

Supplementary Treatment

Many water treatment situations require some sort of supplemental treatment. In general, such treatment will be preliminary to the filter as in the case of ozonation or powdered activated carbon addition. Detailed discussion of such treatment is beyond the scope of this handbook. It should be noted, however, that the optimum diatomaceous earth filter performance should be established before the supplementary treatment tests are initiated. In other words, the grade of DE, body feed rate, and filtration rate that would produce the best results would be determined first. These tests may not produce the required quality water, which would indicate the need for supplementing DE filtration. The supplementary treatment tests would then be conducted using the DE filtration mode established initially. After the desired product water quality is attained with supplementary treatment, it may be necessary to adjust certain filtration features.

Use of ozone may enhance removal of particulates. This improved particulate removal capacity may permit use of higher filtration rates, coarser DE for precoat, or less body feed. To discriminate in a logical manner the nature of cause and effect when treatment changes are made, the filtration adjustments for pretreatment should be made in the same order as before. Start with determination of the precoat grade DE and proceed through the body feed and filtration rate tests.

Test Data

Test information that must be collected includes the physical and the chemical analysis data defining DE filtration as a particulate removal process. In addition, sample analysis data must be generated that demonstrates the effectiveness of the overall treatment train in achieving the established water quality goals. Often, the regulating authorities will require that this later grouping of data be supplemented just to ensure that the finished water quality meets other aspects of water quality not necessarily relevant to a particular treatment situation.

Filter Run Data

These are the data that for the most part define the performance of the DE filter in both its operation and effectiveness. Table 5.7 lists the physical data that must be collected for each test run.

Physical data include filter flow, pressure, the extent of time between pressure readings or to attain certain finished water turbidity levels, and the DE materials used for each run. These are the primary data required to determine DE filtration effectiveness. As indicated in the table, the options available or the adjustments permitted for each filter run include the filter application rate and the type and amount of DE materials used.

Table 5.8 lists the principal analyses made on site that will be required to determine the effectiveness of the filter to achieve the water quality objectives. These may be rules established by the regulatory standards for particulate removal or goals established for a specific test program. It is the practice today to define more stringent finished water quality goals than in the standards, in anticipation of future rule changes. Some data, like pH and alkalinity, do not specifically relate to particulate removal in DE filtration, but are helpful as indicators of changes in raw water quality. Iron and manganese analyses would be made only where removals are accomplished in treatment or where the possibility of infrequent appearance in the raw water exists.

TABLE 5.7. Physical Data per Test Run

*1. Flow: gpm and gpm/sq. ft.
*2. Precoat: type and pounds used per run and lbs/sq. ft. (leaf area)
*3. Body Feed: type and mg/l and lbs per 1,000 gallons treated
4. Time started/time ended: hours per run
5. Pressure Differential: inlet to outlet, start to termination of run
6. dP/dT: average psi of available head used per hour of run
7. Time from start of run:
   - To reach 0.20 treated water turbidity
   - To reach 0.10 treated water turbidity

*Variable to be established for each test run.

TABLE 5.8. Filter Performance: On-Site Sample Analysis

Item		Method or Equipment
Basic	pH	Benchtop Meter
	Turbidity	Continuous Meter
	Color	Color Comparison or Spectrophotometer
	Alkalinity	Titration
Recommended Addition	Particle Count	Benchtop Meter[1]
	Iron[2]	Spectrophotometer
	Manganese[2]	Spectrophotometer

[1]Alternative: manual bench top meter, once an hour

[2]Where relevant

Treatment Train

Supplementary analyses that are usually required by regulatory standards to determine raw and furnished water quality are shown in Table 5.9. These are the same analyses required by the authorities to check the operations of any drinking water treatment plant. As noted in the table, these analyses may have to be supplemented to reflect the requirements of local rules. Except for chlorine demand determinations, analyses for both raw and finished water are the same. Again, the analyses must be relevant to the specific treatment situation. Microscopic examinations, for instance, would be made of surface water samples only, not necessarily for well water sources. Essentially, these analyses define the acceptability of the finished water for human consumptions related to both health and esthetic issues. Some of these data may be developed on-site by the pilot plant operating staff. Others may require analyses in a certified laboratory.

Data Sheet Arrangement

Tables 5.10 and 5.11 are suggested arrangements for data sheets or computer data logging for DE filter pilot test programs. As indicated in Table 5.10, the data includes calculated as well as recorded values. Many times a filter run must be terminated because of equipment breakdown or at the end of the week for a five-day operation. For the arrangement shown, maximum available head loss has been set at 25 psi. If more than 10

TABLE 5.9. Raw and Treated Water: Treatment Train Sample Analysis

Daily	Chlorine Demand SDS (Simulated Distribution System) Chlorine Residual
Weekly	Total Dissolved Solids Total Coliform Count Microscopic Examination (for algae, amorphous matters, etc.) Taste and Odor TOC (Total Organic Carbon)

Notes: 1. Where necessary, certain above analyses to be performed by certified outside laboratory. 2. Sample items to be supplemented by water quality requirements of overall treatment train and as required by regulatory authorities.

hours' successful run has been completed at termination, calculations may be made to project the time and production values to end of run, which is when a 25 psi differential may be reached.

The data collection format must reflect the particular treatment conditions. For instance, Table 5.10 indicates preozonation that might not be pertinent for certain treatment applications. As shown on Table 5.8, instrumentation may be available to continually record information like head loss, turbidity, or particle count. If continuous monitoring is not available, data must be collected by observation. Turbidity or particle count values might have to be taken every few minutes until reasonable steady state condition exists. Thereafter, data would be recorded once an hour. Where multiple data observations must be made, separate data sheets might be advisable. Table 5.10 is a summary sheet reflecting changes from start to end of run for successive runs.

Table 5.11 shows data that may not be typical for all treatment situations because of raw water quality or availability of equipment. Particle count is becoming more important in water treatment data recording because of the need to monitor removal of particles in the 3 to 15 micron size range. This range includes the typical sizes of the *Giardia* and *Cryptosporidium* cysts.

Particle count is important because it indicates the true value of DE filtration in removing particulate matter. Turbidity is only an indicator. It has been demonstrated that DE filtration achieves particle removals to lower levels than alternative treatment where finished water turbidity may be the same.

Total organic carbon (TOC) provides a measure of the susceptibility of treated water to generate bacterial regrowth in a distribution system as

TABLE 5.10. Typical DE Pilot Test Data Schedule

ITEM\RUN NO.	1	2	3
OPERATING DATA			
Start			
Date			
Time			
Pressure Differential, psi			
O_3 Rate, mg/l			
DE Material			
Type			
Precoat, lbs			
Body Feed, mg/l			
Filter Rate, gpm/sq. ft. (calculated)			
Turbidity/Time			
Start Run, NTU			
Time to 0.2 NTU, minutes			
Time to 0.1 NTU, minutes			
Steady State NTU			
Color (steady state) CU			
End of Run			
Date			
Time			
Pressure Differential, psi			
Run Length, hours			
Pressure Rise, psi			
CALCULATED			
dp/dt, psi/hour			
Project Time, hours (to 25 psi)			
Total Production, gallons			
Actual			
Projected (to 25 psi)			

TABLE 5.11. Supplementary DE Test Data

ITEM\RUN NO.	1	2	3
Iron			
Manganese			
Particle Count, No.*			
2–3 μm			
3–5 μm			
5–8 μm			
8–15 μm			
>15 μm			
Tot 3–15 μm			
TOC, mg/l			

Note: All above data to be taken after test run has reached steady-state turbidity condition.

*Particle count grouping may vary according to test requirements or instrument used.

well as the THM formation potential. In any case, the format for data collection either manually or in computer data logging must reflect the requirements of the particular pilot program.

Principal Procedure Considerations

There are many facets of planning and operating a DE pilot test program. The principal test determinations that should be included in any pilot investigation are shown in Table 5.12. This is a minimum listing that may require additions depending on the particular treatment situation. Details of the basic considerations listed in the table are discussed later.

Diatomaceous Earth Filtration Viability

Any program should start with the relatively inexpensive bench top tests described before to determine whether DE filtration is capable of producing finished water close to the quality defined in the rules. If possible, these tests should be conducted with raw water samples with the highest turbidity or close to highest turbidity levels to be expected (indicated by

TABLE 5.12. Principal Pilot Program Test Determinations

1.	Viability of DE filtration
2.	Type and use of DE for precoat and body feed
3.	Maximum filter application rate
4.	Need for supplemental color removal pretreatment
5.	Other supplementary pretreatment and post-treatment
6.	Performance over changing raw water quality conditions
7.	Potential for intermittent pilot operation.

past records). A pilot program may be a costly commitment that might be wasted if the current raw water quality is superior to that experienced on the average.

Once the viability of DE filtration is established, the pilot program is designed. The program is more than the treatment hardware. The program should include providing sufficient space for the pilot treatment train, laboratory test accommodations, desk space for operating personnel, developing test procedures, test schedules, data sheets, and possibly computer availability for data logging. Once the program has been designed and the pilot equipment and other accommodations installed, the tests can proceed following the subsequent steps listed in Table 5.12.

Diatomaceous Earth Use

The first step in the test program would include determination of the body feed grade. In general, Hyflo, produced by the Celite Corp., or its equivalent, produced by Eagle Picher, are the most common grades of DE used in drinking water treatment. Usually coarser grades of DE than these would not be used. It may be necessary, however, to use a finer grade for some applications, particularly where colloidal particulates may be present. An alternative to finer grades would be the use of alum coated media.

Normally, precoat would be applied to the filter close to the amount of 0.20 lbs/sq. ft. of filter leaf. While this amount seldom varies, some situations may require a heavier precoat.

Most of the initial tests would be devoted to determining the most efficient use of body feed. The effect of varying the rate of body feed addition was discussed in Chapter 3; illustrated in Figure 3-3. The most efficient use of body feed is attained when head-loss vs. time reaches a straight line plot with the minimum body feed addition. Once this is determined, the body feed addition should remain constant until there is a change in raw water quality affecting filter cake quality. This feature of DE operation makes it possible to project run, time, and production values for incomplete runs. In any case, complete run data should be plotted periodically to ensure the straight line characteristic. Even if raw water quality data does not indicate significant change, body feed should be reduced at times to ensure efficient use.

Maximum Filter Rate

The maximum possible filter rate must be determined for both "worst case" and treatment conditions at other times, usually for periods in which there is little change in source water quality. The "worst case" operating conditions will define the design capacity of the full-scale plant. It is somewhat important to establish the type of operation of the full-scale plant to determine the maximum feasible rate. If it is known that the ultimate plant will provide operator attendance only for one shift a day, the filter runs and cleaning should not take place more frequently than once every 24 hours. If the plant is attended around the clock, then a worst case run time of 12 hours might be feasible. It has been found that a cleaning frequency of more than twice a day is not practical. Precoat use, time out of service, and operator attention would be excessive at greater cleaning frequencies.

It is also important to know the maximum filter rates possible at times other than worst-case. DE filters may easily be taken out of service and placed on standby. Evaluation may show cost savings realized with fewer DE filter units in service at higher rates for extended periods when the source water quality permits.

Fortunately, it is not critical that filter rate determinations be conducted under precise raw water quality conditions. As long as the body feed rate is close to maximum efficiency, production rates (application and volume) may be accurately calculated for filter application rates other than those tested. This is discussed in Chapter 3 and illustrated in Figure 3-3.

Need for Supplemental Color Treatment

Color may be present in raw water either in dissolved or colloidal form or both. Supplemental treatment may be required to attain less than the 5 CU maximum limit. If the problem is predominantly colloidal color, using alum coated DE for part of the precoat will successfully remove the normally negatively charged colloids. The use of a strong oxidant like ozone or chlorine will break the bonds of organic matter with the color particulates and effect neutralization of the negative charges. The color colloids will then be removed through attachment in the DE precoat and in the cake accumulation. This is also discussed in Chapter 3. Oxidation can also accomplish precipitation and/or bleaching of the dissolved color.

The final disinfection of a drinking water before introduction into a distribution system will also accomplish some bleaching of color. In many instances, therefore, where DE filtration without supplemental treatment accomplishes color reduction close to, but not below 5 CU, final disinfection may produce acceptable color levels. This possibility should be determined by tests.

Other Supplemental Treatment

Detailed descriptions of supplementary treatment pilot tests other than those discussed previously are beyond the scope of this text. Most of the more common pre-treatment and post-treatment processes that may have to be considered for different applications have been reviewed in Chapter 3 and Chapter 4. Supplementary treatment in DE pilot filter trains should simulate the full-scale process and, as much as is possible, be capable of operation at the same rate of flow as the filter unit. Many pretreatment pilot units will have to be operated at higher rates so excess flow must be wasted. Some post-treatment tests like those for chlorine demand and chlorine residual may better be accomplished with bench top determinations.

Changing Raw Water Quality

Once a pilot program has started, it is important that the treatment train be tested over varying source water quality conditions that may affect filter performance. These conditions may include periods of adverse water quality such as reservoir bottom turnover and excessive rainfall causing washdown of terrestrial detritus to the source water. Tests also should be conducted for seasons of good water quality to determine the possibility of operating few filters at higher rates as discussed previously. Most important, these operations during varying conditions of raw water quality will establish a meaningful basis for calculating the annual amount and costs of DE used for precoat and body feed and the power costs for filter supply pumps. These two features together comprise a significant proportion of the total annual costs for DE filtration plant operation.

Intermittent Operation

An important advantage of DE filtration is the fact that a filter may be taken out of service for extended periods without impact on treatment. This ability to "moth-ball" a DE filter also makes it possible to run a pilot test program on an intermittent basis. The test program may be halted for long periods of time when there is little change in source water quality. Alternatively, it would also be possible to operate the pilot filter one day a week when raw water quality is static. This latter procedure is probably superior because it does permit obtaining long-term data with minimum operating time.

Pilot Operation—Troubleshooting

The small scale of a pilot DE system demands that equipment be maintained in optimal operating condition and that scrupulous attention be given to control of the operation and collection of data. Depending on the nature of the system fault or data error, even minor deviations from actual operating conditions may be magnified many times.

Pilot plant operation may be divided into two general areas of concern.

1. Maintaining the equipment in the system to ensure reliable functioning of the individual components
2. Assuming reliable equipment function, evaluation of results, and troubleshooting the features of operation that must be varied and/or controlled to respond to raw water quality changes, namely:
   - Filter flow or application rate
   - Precoating
   - Body feed addition
   - Valving (flow routing) changes by cycle or treatment step.

Mechanical Maintenance

Filter leaves must be kept clean and filter leaf-collection manifold seals must be absolute for proper operation of pilot filter. A clogged screen will result in incomplete precoating, short filter runs, and possible turbidity breakthrough. If the leaf-connection seal is not tight, part of the filter inlet flow may leak through, resulting in breakthrough and loss of product clarity. Clogging may be caused by fine DE not removed in cleaning and embedded in the screen. Some features of raw water quality like iron and manganese may coat or scale the screen mesh to also cause clogging. Since it is inert, embedded DE must be removed by high-pressure sprays directed at the leaf that is removed from the filter. There are various chemical components that are commonly used to dissolve coated iron and manganese, the principal causes of the scaling problem.

Most leaf-connection seals are of the "O" ring type that may be easily replaced. Special care must be exercised when placing the filter leaf into the manifold pocket to avoid damage to the seal. Damage rather than wear is the principal cause of seal leakage.

The other principal mechanical faults that may be encountered are related to accuracy and reliability of metering pumps. Abrasive DE can wear meter pump checks, especially those of the ball variety. Metering pumps can also become clogged. This happens mostly because a meter pump operation is halted for a long period of time, leaving DE slurry in the pump. Meter pump accuracy may be easily determined by checking the drop in slurry volume in the suction tank. Plunger/diaphragm type pump heads should be removed and cleaned periodically. Peristaltic pump hoses should be inspected for condition.

Filter Flow—Application Rate

During the precoat cycle, the application rate must be low enough to produce a uniform thickness of coat across the leaf. In filtering, the rate should always produce more than 12-hour filter runs. Both rates should be high enough to generate sufficient hydraulic force to hold the precoat or accumulated cake against the screen. In general, a rate of 0.25 gpm/sq. ft. is sufficient to hold precoat or accumulated cake in place. Principal problems encountered with improper flow rates and the identification of those problems are as follows:

— Too low flow during precoat may permit DE to settle in the filter, resulting in insufficient DE left to precoat to the proper thickness, which can allow turbidity or color breakthrough.

- Too high an application rate may wash precoat or prevent coating of those parts of the leaf in the more turbulent flow areas, resulting in spotted, non-uniform precoat, which may cause turbidity breakthrough.
- Too high a rate during filtering can shorten filter runs to less than 12 hours.
- Too high a filter rate for vacuum filters, particularly, can permit dissolved air in the incoming flow to be released in the higher velocity flow generated through the accumulated cake interstices. This can "pop" the cake and its bridging across the screen openings resulting in breakthrough of DE particles.
- Too low a flow rate during filtering or interruption of flow may cause the cake to separate from the septum or to crack due to too low a holding force, which can permit turbidity breakthrough.

Precoating

This initial operation should produce a uniform precoat of about 1/8 inch thick. After the precoat DE has been added, the recycle flow should clear up significantly in two to five minutes' time. After a total of recycle time of 10 to 15 minutes, full clarity should be reached and the flow mode is switched to filtering. The principal problems encountered may be caused by the following:

- Too coarse grade of DE or insufficient DE to produce the proper coat thickness may permit breakthrough and the inability to reach 0.20 NTU turbidity.
- Too low a flow rate during filtering or interruption of flow may cause the cake to separate from the septum or to crack due to too low a holding force, which can permit turbidity breakthrough.
- Excessive recycle time for precoating may eventually drive the finer DE particles through the precoat layer and leaf mesh to eventually recycle and deposit on the face of the precoat to cause "nesting," lowered porosity, and shortened filter runs in the next cycle.
- If either a pressure or vacuum pilot filter system is not completely filled at the start of the precoat cycle, less precoat will be deposited at the top of filter leaves, providing an area for potential breakthrough during the filter cycle.

Body Feed

Body feed should be added at the minimum rate required to produce a straight line plot of accumulated head loss vs. time during a filter run.

— Too high or too low a body feed addition rate will shorten filter runs. Too high a rate can also cause bridging of cake between leaves of multiple leaf pilot filters to shorten runs.
— Interruption of body feed addition during filtering will result in clogging of the cake in that period, which will shorten filter runs.

Valving—Flow Routing

The principal valving changes are those to place the DE filter system in the precoat mode of flow, the filtering mode, and in a drain arrangement for cleaning. Only the change from the precoat to the filtering cycle requires special attention to ensure that the recirculation valve is closed simultaneously with the opening of the raw water supply inlet. Failure to accomplish the simultaneous switchover can result in interrupted flow, which may affect cake integrity.

6
Equipment and Materials of Construction: Principal Safety and Design Features

Influencing Factors

Except for the filter unit itself, the components and facilities of a diatomaceous earth filtration plant are similar to those in more traditional water treatment installations. Dry materials may be delivered in bags stacked on wood pallets. Consistent with the trend, chemicals for post treatment are usually in liquid suspensions, stored in tanks of plastic construction fitted with mixers (if necessary). For the most part, shut off and control valves will be of the butterfly, gate, or ball types. The differences in basic safety and design considerations for a DE filtration plant are related more to the diatomaceous earth material itself. Unlike materials and chemicals used in more traditional water treatment, diatomaceous earth is a siliceous, abrasive material that requires certain precautions in both plant design and operation. On the plus side, diatomaceous earth is an inert, non-corrosive material and does not require the special design consideration of most chemicals and materials used in other traditional processes.

Dry Media

As a fine powdered abrasive siliceous material, dry diatomaceous earth may cause irritation of the eyes and respiratory systems of plant personnel if incorrectly handled. Containing cystalline silica, DE has been classified as a potential source of cancer if inhaled. Compared to other more hazardous water treatment materials like hydrofluosilicic acid, caustic soda, lime, and chlorine, handling dry DE requires simple precautions and care to avoid harmful effects.

As mentioned before, dust collectors must be provided wherever dry DE is transferred from bags to hoppers or slurry tanks. As specified in the AWWA Manual on Precoat Filtration:

> Personnel handling the filter media should be supplied with appropriate respirators, gloves and goggles, and an eyewash should be available. Personnel should be cautioned not to rub their eyes and to use the eyewash in the case of irritation. Coveralls or rubber aprons for operators are recommended. Safety precautions for a specific product should be followed according to the material safety data sheets provided by the supplier.

Wet and dry vacuum pick-up equipment should be available at all dry storage and transfer locations where spills may occur due to broken bags or careless handling. As a word of caution, where spilled DE may be collected in floor drainage systems, sufficient cleanouts and flushing connections should be provided to avoid clogging.

In slurry operations, dry media should be mixed into moving water, with such movement provided by the turbulence of water entry, mixers in tanks, and eductors in pipelines. Rooms for storage of dry materials should be well-ventilated with all exhausts into the atmosphere provided with some type of dust separation device.

Diatomaceous Earth Slurries

Design of provisions for handling and conveying DE depend on the concentration of the slurry suspension. While it is somewhat arbitrary, the division between light and concentrated slurry may be established at an 8% suspension. Experience has shown that there is little wear of conveyance and storage facilities at lower concentration. There has been noticeable wear with slurry concentrations between 12 and 20%.

Conventional water treatment facility design may safely be used for concentrations below 8% with a minimum of special provisions. Despite low concentration, DE particles do tend to settle and collect in "dead" areas so that some simple precautions may be beneficial in design.

Design for slurry concentrations above 8% requires recognition of the "wear" possibilities, especially where there are narrowed flow areas, changing flow directions, and higher generated turbulences like in centrifugal pumps. Special care must be exercised in the selection of most pipelines, valves, and fittings as well as equipment. These special accommodations are discussed later by facility.

Special Design

It is not the intent to cover the design specifications of all aspects of DE filtration facilities in this text. Many of the facilities would be identical to those in more traditional plants for which more than adequate design detail information can be found elsewhere. The facilities discussed later are those that require special design provisions and conditions because of the unique nature of the fine, abrasive, siliceous DE material. Many standard items of equipment available from manufacturers are suitable for DE design applications. While this equipment may have been designed for handling other types of materials, there can be close association with the wear characteristics of DE.

System Hydraulics

DE system hydraulics must be adjusted for the amount or concentration of DE suspension. The need to control the velocity of flow of DE suspensions was discussed previously. Too low a velocity will permit DE particles to settle in pipelines to inhibit flow; too high velocity may cause pipe wear. In general, the lowest velocity that will keep suspended material mixed in the stream of flow will be the most effective. This will result in the minimum pressure drop due to friction, the least abrasion of pipe walls, and the least attrition of the somewhat friable DE. While lower velocities would be permissible because of the extremely slow settling rate of DE particles, minimum pipeline velocities of 1.5 to 2.0 ft/sec would provide for practical design. As discussed in Chapter 4, maximum velocities should be limited to less than 9 fps for concentrations less than 8%, 5.5 fps for concentrations more than 8%.

Pipeline pressure loss is increased over that for clean water in proportion to the concentration of DE particles. Tests have indicated that a reasonable adjustment for pressure loss may be achieved by simply multiplying the clean water loss by the specific gravity of the suspension. Using a specific gravity of wet Hyflo of 2.3, the net specific gravity of various concentrations of DE suspensions were calculated. For ease of use, these have been plotted for a range of concentrations up to 20% in Figure 6-1.

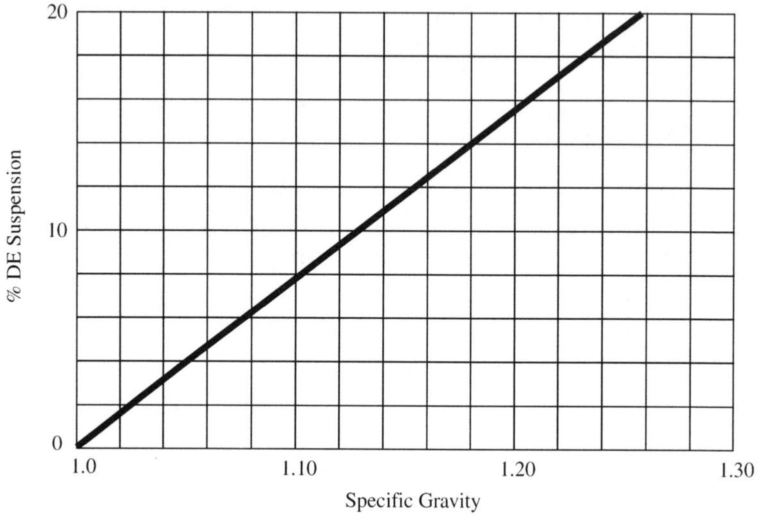

FIGURE 6-1. Specific Gravity Diatomaceous Earth Suspensions

An example of the calculation is as follows:

— Clean water pipeline head loss: 6.7 ft
— Specific gravity 14% suspension: 1.18 (from Figure 6-1)
— Adjusted head loss: 6.7 ft × 1.18 ft = 7.9 ft

As emphasized before, DE suspensions must be kept in motion to avoid settling and eventual clogging of pipes and vessels. Once flow is interrupted for more than a few minutes, the pipeline should be flushed clean, for which accommodations should be provided as discussed in Chapter 4.

Water Supply Pumps

Water supply pumps in DE filtration systems handle either the raw water supply to the filters or serve as recirculation pumps for precoating. In many systems, one pump might serve both purposes. For all practical purposes, these pumps would be the same used for all-around water supply service. The most common type probably would be the split case, double suction unit, which has been the work horse in the water industry for years. Pumps would be selected that would provide the most efficient operation performance over the range of hydraulic conditions. Those conditions most pertinent to DE filtration service would be the increase in pressure requirements from start to end of DE filter run. This may range from 20 to 35 psi increase for pressure filters and a reduction in available suction pressure of from 10 to 18 feet in vacuum filter systems.

Pump shaft seals may be the packed or mechanical type. Where the pump may provide recirculation or dual service, packed seals must be provided with sealing water to avoid DE particles damaging the pump shaft.

Pump selection features for both pressure and vacuum type DE filtration are discussed later. In general, the primary objective would be to select the pump that would provide the most efficient operation (least power cost) while satisfying all other performance requirements.

Pressure Systems

The notable feature of pressure DE filter operation is the rather large increase in head loss across the filter from start to end of run. A pressure increase of from 20 to 35 psi could constitute from 50 to almost 80% of the required total dynamic head (TDH) rating of the supply pump. With such a significant change in head, a throttling valve is usually installed with a flow measuring device control in the discharge line from the filter to maintain constant flow as the head loss increases. This means that almost half the electric power expended from start to end of run is for all practical purposes wasted.

An important alternative to the automatically throttled valve for flow control is control of the rotating speed of the pump. The required TDH or the head delivered by the pump will then vary to compensate for the gradually increasing head across the accumulating cake. The relationship of increasing pump speed to deliver increasing head is expressed as follows:

$$\frac{H1}{H2} = \frac{(N1)^2}{(N2)^2}$$

Where:

H1 = head @ N1 rpm
H2 = head @ N2 rpm

Horsepower requirements change with rpm as follows:

$$\frac{Bhp1}{Bhp2} = \frac{(N1)^3}{(N2)^3}$$

Curves have been developed from these pump rotation relationships to illustrate potential power savings with variable speed pumps. As an example, assume that a pump delivering a fixed flow rate must develop a head increase of from 10 psi (clean filter) to 35 psi (end of run). The initial head loss ratio may be computed as a percent of the total dynamic head (TDH) at maximum speed as follows:

$$\text{Initial H} = \left(\frac{10}{35}\right) \times 100 = 28.5$$

From Figure 6-2, the pump would operate at less than 20% of the horsepower required under maximum pumping conditions. Pump rpm is calculated to be around 53% of maximum. At 70% of the developed head, the required horsepower would be around 57% of maximum with pump

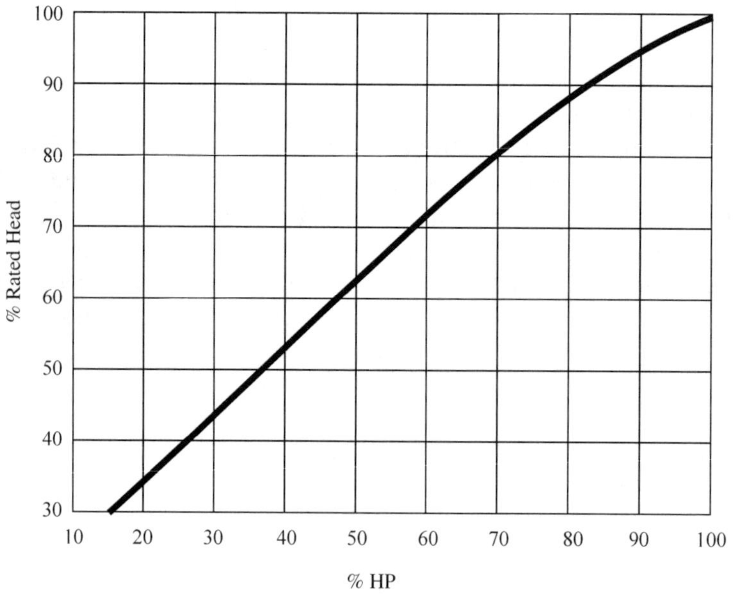

FIGURE 6-2. Percent Rated Head vs. Percent HP

rotation at 83% of maximum. It is obvious that considerable power savings may be realized by controlling pump rotation rather than excess head to provide for the increase in head requirements as the filter cake accumulates.

The automatic speed control could also adjust the pump flow for precoat service where the supply pump is used for dual service.

Vacuum System

The single feature that distinguishes vacuum DE filter operation is the suction of the supply pump providing the driving force for flow through the filter. Actually, the available barometric pressure provides the driving force. The available differential head is equal to the barometric pressure less the sum of the following:

- — Vapor pressure of water at operating temperature
- — The net positive suction head (NPSH) required by pump to maintain delivery of the required flow to the pump impeller
- — Clean system (filter leaves, manifold, and piping to pump) head loss.

The relationships are shown in Figure 6-3. As indicated, barometric pressure is determined by the altitude of the site and the temperature of the water; vapor pressure by the temperature of the water being pumped. The NPSH requirement varies according to the pump design and its value is available from the particular manufacturer. For example, the site variables for a DE vacuum filter plant located at an altitude of 1,000 feet above sea level pumping 70 °F water are as follows:

- — Barometric Pressure = 32.8 ft of water
- — Vapor Pressure (70 °F water) = 0.89 ft
- — Specific Gravity (70 °F water) = 0.998
- — Adjusted Barometric Pressure = (32.8/0.998) = 32.9

As indicated, the adjustment of barometric pressure in the range of temperatures for drinking water sources is negligible and may be disregarded. Vapor pressure of water at 80 °F increases to 1.2 feet of water and decreases to 0.59 feet at 60 °F. It would, therefore, be adequate to assume a nominal 1.0 feet vapor pressure in all vacuum filter calculations. It follows that the barometric pressure, altitude, clean system (suction) head

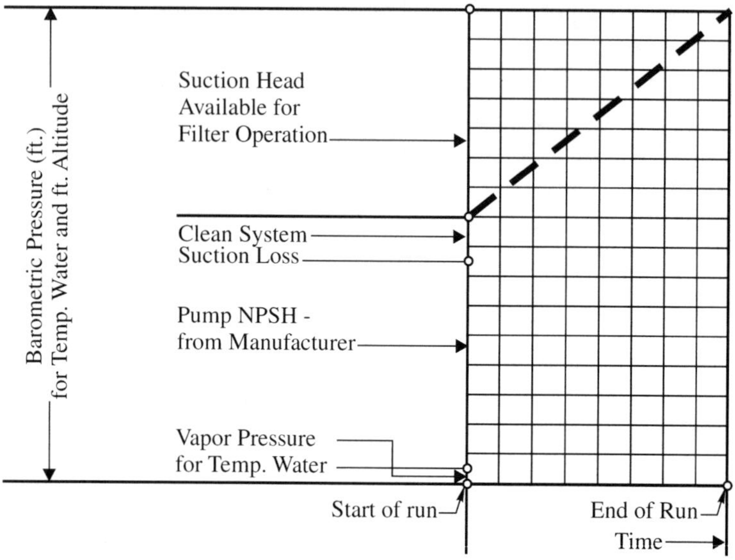

FIGURE 6-3. Schematic, Available Section Head, Vacuum Filter Supply Pump

loss, and pump NPSH are the significant suction hydraulic issues. The NPSH is by far the most significant factor so that in vacuum system design it is necessary to specify pumps with minimum NPSH requirements to attain the highest available head for vacuum filter operation. In general, available head should vary from 12.0 feet up to 20 feet (for the most efficient pump design).

Throttling valves modulated by controls paced by a flow measuring meter installed in the supply pump discharge line would always be used for flow control in a vacuum filter system. The relatively low available head would make pump speed control impractical.

Slurry Handling

For these discussions, slurry is considered to be concentrated, requiring special design features for DE suspensions of more than 8%. Regardless of the mass of DE in suspension, all slurries in storage require mixers to prevent settling. The mixers would be specified to maintain sufficient vertical velocity and movement to maintain a homogeneous mixture to ensure accurate feed rates of DE slurry to process.

Mixer Design

Over time, DE particles in suspension tend to break down in size. To inhibit such attrition, excess energy over that required for suspension should be kept to a minimum and, except for storage of DE filter wastes, storage time should be limited to no more than 3 days. The functional aspects or objectives of mixing should be specified by the process designer. The proper mixer should be selected and designed by the manufacturer. The attrition of DE particle size is more evident at higher concentrations where more care must be exercised in providing both mixer energy and storage detention before use.

Tanks

Despite the abrasive qualities of diatomaceous earth, plastic tanks provide more than satisfactory service for storage of DE slurries, even those of high concentration. Storage tanks may be divided into two categories.

— *Process Service:* providing storage of slurries added directly to process for precoat and body feed
— *Bulk Storage:* for initial preparation of slurries that are subsequently distributed to the process service tanks.

More accurate control of DE slurry feed to process may be achieved with suspensions of lower DE content. Generally, suspensions used for precoat and body feed are maintained at from 2 to 5% DE. Bulk storage concentrations are more than 12% DE with variations usually up to 18%. This range of concentrations allows the designer some choice in the selection of storage tank size and slurry suspension distribution rates.

The type of plastic construction specified and the tank support depends on the size of the storage tank and the cost. The largest storage tanks are constructed of fiberglass reinforced polyester; the smallest tanks usually are formed of unreinforced polypropylene or polyethylene. Thin-walled, lightweight small tanks may be used if provided with a steel tank pack or container. Polypropylene tanks with a relatively snug fit in a recycled 55-gallon steel drum are common for this latter arrangement.

Large tanks are flat-bottomed, installed directly on the floor so the floor rather than the tank bottom supports the liquid load. Smaller tanks may be installed on fabricated steel platforms, elevated above the floor to provide for more convenient piping connections. Because DE is inert in character, slurry tanks do not require the containments specified for many chemical solutions in the regulations. It would be prudent, however, to provide containments for areas where multiple bulk storage tanks may be installed. While not harmful, DE slurry spilled over large areas is not easy to pick up.

Waste slurry discharged in the cleaning of DE filters may be stored in plastic tanks or reinforced concrete basins. Many times, the waste treatment process will require thickening or decanting to reduce storage volume or to facilitate dewatering. Concrete tanks can provide improved structural support over lighter-weight plastic construction for this heavier piping and equipment.

Mixers are mounted directly on slurry tanks where the structural strength of the tank permits. Where smaller tanks are installed on steel support stands, the mixer can be clamped on a steel side member extended vertically from the stand. Heavy mixers for larger plastic tanks require special structural support so the mixer unit may be installed directly in the center of the tank. A typical arrangement for large tanks is for the mixer motor to be mounted on the floor above the bulk storage tanks. Mixers for concrete basins are mounted on a steel support across the top, which would also provide walkway access to the gear motor drive.

Slurry Transfer

Where slurry is prepared remote from the point of use, transfer pumps must be provided. It is noted that for small-capacity DE filtration systems, 50-lb bags of DE or a weighed amount of DE could be emptied directly into the service tanks for slurry preparation. The usual dust collection precautions must be observed in these installations. For larger systems, the slurry may be prepared first in bulk storage tanks at higher concentrations, then pumped to the service tanks where dilution water is added.

Several types of pumps are available for pumping concentrated DE slurry. These pumps have been designed and used for pumping other suspended, abrasive material for decades. The three principal types of pumps that might be used for transfer of concentrated slurries include:

— End suction centrifugal pumps manufactured with thick, case-hardened volutes and impellers designed for slurry service with back vanes to wipe solids out of the stuffing box area. These pumps may be fitted with replaceable end suction liners, which take most of the wear on the suction side of the pump casing.

— End suction centrifugal pumps designed with the casing split vertically normal to the shaft alignment. Oversized casings would permit installation of two-piece rubber liners held in place by the flanges connecting the two parts of the casings. The rubber liners take the wear. Special abrasion-resistant impellers would be used.
— Peristaltic or hose pumps now available in larger capacities. Rotating vanes, compressing the hose against an outside cylinder, literally "squeeze out" the slurry in a continuous action. The advantage of this pump is that there are no moving parts in direct contact with the abrasive slurry. An easily replaced hose takes the stress and wear of the pumping action.

The above pumps would be used only for concentrated slurry service. Standard water service pumps may safely be used for pumping where the DE suspension is less than 8% DE by weight.

Metering Pumps

The most important feature of metering pumps used to feed body feed to process is accuracy. Fortunately, body feed is added to process most often in suspensions of DE of from 2 to 5%. Two types of metering pumps are commonly used for body feed service.

— The plunger diaphragm pump is the most commonly used in all water treatment chemical solution feed applications. These pumps may be fitted with easily replaced corrosion- or erosion-resistant ball checks and seals at the inlet and outlet. The rate of feed is changed by adjusting the length of plunger stroke and/or the speed of rotation of the pump, which controls the plunger cycles per minute.
— High-accuracy peristaltic metering pumps where the adjustable speed of rotation of the compression vanes provides the means for changing the rate of feed.

In operation, the pump rate adjustments may provide the means for increasing and reducing the rate of slurry feed proportionately; the base rate is determined by the volume of slurry pumped from the suction tank over a period of time. Because of the abrasive nature of DE slurry, this base rate should be checked at least weekly for a continuous filtration operation by measuring the drop in the suction tank over time.

Pipes, Valves, and Fittings

Conventional waterworks piping practice may be followed for piping in a DE filtration system where the DE concentration is less than 8%. Principal pipe materials used would include cement-lined ductile iron pipe, polyvinyl chloride (PVC) plastic pipe, and the several available options of steel pipe. Even with low concentrations of DE, the design should provide for minimum carrying velocities to inhibit settling of DE particles. Valves and fittings for water piping and pipes that may carry suspensions of less than 8% DE would also be in accordance with standard waterworks practice.

Recommended piping practice for systems carrying slurries more than 8% in concentrations are discussed here.

Piping

Cement-lined cast iron pipe would not be used for carrying concentrated DE slurry. The abrasive DE would eventually wear away the cement lining, especially at bends and fittings. The two types of pipe best adapted to handling slurry include:

- Relatively inexpensive, easy-to-install PVC pipe. Schedule 80 pipe should always be specified. While this pipe may eventually wear, the thickness will allow many years' service before replacement is necessary.
- Thin-walled stainless steel pipe. Since there would be no corrosion problems, the least expensive grade of stainless steel would be adequate. Developments in thin-wall construction have reduced the cost of this pipe both in materials and the installation of the reduced-weight pipe. Stainless steel would be one of the more abrasion-resistant type of piping available.

Where available, long radius bends should be used in the piping arrangement. Use of tees should be kept to a minimum.

Valves

Globe valves should not be used for concentrated DE service. Gate valves (solid wedge construction), butterfly, and ball valves may be employed, but unless pipelines are kept flushed clean after use, trouble may be expected with increased wear on the moving parts. Perhaps the least troublesome of these three valves in slurry service would be the ball valve. Valves constructed of stainless steel may be obtained to reduce wear.

The pinch valve is the best for concentrated slurry service. This valve is simply a rubber sleeve with flanged ends fitted into a steel frame. The frame has a threaded shaft and turning handle that lowers a metal bar across the sleeve for both throttling and shut-off. This valve will provide the most reliable slurry service. The valve may be removed for relatively quick replacement of a worn sleeve.

Instrumentation and Controls

Instrumentation and controls used for DE filtration systems are no different in concept than those used in traditional water treatment plants. Because of the DE particulates carried in the flow lines and their abrasive nature, precautions should be observed as follows:

- — Inserts such as venturi sections or other type probes should be stainless steel rather than plastic.
- — Ports into pipes for sample connections for measuring turbidity and color as well as for flow meter differential connections should have provisions to permit flushing back into the pipe.
- — Where necessary to measure the flow of concentrated slurry, magnetic or ultrasonic flow meters should be considered where no penetrations of the pipe are necessary.

The simplicity of operation of the DE filtration system lends itself to computer-controlled operation of pressure filters which may be automatic or partially automatic. The three cycles would be controlled in the sequence previously indicated in Table 4.1 and as described in more detail later.

Precoating

- — After a cleaning cycle has been completed, the computer controls actuate the appropriate valves to place the piping in the recycle mode.
- — A motor- or solenoid-operated control valve is then actuated by a timer to fill the filter and system with filtered water. The valve is shut by a level control once the proper system level is reached.

- When the inlet water valve is shut, the recirculation pump is started with the flow rate measured by a meter that controls the desired rate by adjusting a throttling valve or by varying the pump rotation speed. This is followed by introduction of a fixed volume of precoat slurry by pump, which is stopped by a timer switch. The pump rate, time, and slurry concentration define the volume added.
- A turbidity meter connected to the filter outlet line measures turbidity of the recirculated flow. Once this flow runs clear (turbidity under 0.5 NTU), the valve positions are simultaneously changed to switch from the recirculation to filtering mode. The same turbidity meter also monitors filtered water turbidity.

Filtering

- When the switch to the filter flow mode is made, the pressure differential meter with connections to inlet and outlet of filter is actuated.
- The flow rate through the filter remains at a constant preset rate controlled by means of an effluent line throttling valve or by varying the pump speed of rotation.
- With the assumption that the clean filter system differential pressure is reasonably constant, the filtering mode is halted when a preset pressure differential is reached.
- The pump is stopped and the valve positions are changed from the filtering flow pattern to that for cleaning where the filter drains to waste.

Cleaning

- First the drain valve opens and the filter is completely drained. When the filter is completely empty, indicated by a timer or hydraulic probe, solenoid valves open to provide filtered water for the motor-operated rotating spray nozzle headers.
- The length of the cleaning cycle is controlled by a timer. When the preset time has elapsed, the flushing water supply is halted and the filter valving is changed back to the precoat position.

Alarms

With automatic operation, it is necessary to use alarms to indicate system fault. Some basic alarms that may be considered include:

— *Precoat.* If after a preset elapsed time, the recirculation flow has not reached the required turbidity level, an alarm sounds and the recirculation pump stops holding the valves in the precoat recirculation mode.
— *Filtering.* If the desired finished water turbidity is not achieved after a preset time, the alarm sounds and the supply pump is halted with valves held in the filtering mode.
— *Cleaning.* A pressure switch is installed in the flushing water lines. If sufficient pressure is not indicated before the spray manifolds are actuated, an alarm sounds and the cleaning cycle stops. If the supply of flushing water and attendant pressure are lost in the middle of the cleaning operation, the alarm sounds and the spray manifold rotation stops.

If the precoat or filtering operation is stopped for any reason, the filter is switched (manually) to the cleaning mode to start over again.

The automatic computer-controlled feature may be partially used as follows:

— After cleaning, the precoat step is initiated manually with automatic control through the next cleaning. This allows the operator to inspect the interior of the filter through receiving ports to ensure a clean filter before initiating a new run.
— The automatic feature for cleaning is used only after the filter run is halted by loss of head.

Automatic computer control of the DE pressure filter is an ideal operating arrangement. The same control systems may be used for a vacuum system fitted with rotating spray manifolds for cleaning and tightly sealed covers to contain the spray and spray rebound. If the covers are installed, however, one of the major benefits of the vacuum filter system would be lost: the ability to inspect the filter through all of its operating cycles. In addition, manual cleaning of the filter leaves by high-pressure hose is an excellent way to ensure clean leaves before startup.

7
Design and Operation

Most of the basic information required for the design and operation of diatomaceous earth filtration systems has been reviewed in the previous chapters. The purpose of this chapter is to present, in general terms, the preparation required before starting a design and the sequential steps that would be followed in the design process. This is followed by a guide for troubleshooting filter operation problems both in initial startup and thereafter.

Before design proceeds, it must first be demonstrated that DE filtration would provide the more advantageous treatment for a particular drinking water source. Some beneficial applications of DE filtration would be more obvious than others. In any case, with the constantly changing drinking water treatment regulations, both in stringency and broadening scope, it will be necessary to conduct a test program to demonstrate the effectiveness of DE filtration. In some instances, it may also be necessary to conduct parallel tests with more traditional rapid sand filtration treatment trains to demonstrate the superiority of DE filtration. Drinking water treatment, its regulations, and academic research efforts, have virtually been confined to the rapid sand filter process and its variations for almost a century. A well-organized and well-executed process investigation program is, therefore, essential to ensure acceptance of DE filtration where it is demonstrated to be the more beneficial alternative.

Potential Applications

Water Quality

In evaluating the use of DE filtration in a particular treatment situation, both its benefits and limitations must be considered. Without supplemental treatment, DE filtration is strictly a particulate removal process. Since it provides for one-step solids removal, DE application will generally be limited to treating source water with turbidity 5.0 NTU and below. In unusual circumstances, treatment of higher levels may be possible where turbidity may be contributed by silt-like solids partially within the size gradations of the DE media used.

Color is difficult to remove in DE filtration unless it is in colloidal form where natural charges may be neutralized with the use of strong oxidants like ozone, chlorine, and chlorine dioxide. Final disinfection by chlorine can "bleach out" part of the color. The overall treatment train color reduction potential, therefore, must be investigated.

With an increasing regulatory emphasis on DBP (disinfection byproduct) levels, it is evident that DE filtration would be limited to waters of lower organic content. This problem, however, may be overcome by the use of alternative disinfectants like chloramines that do not react with NOM (natural organic matter) to form DBPs. In many applications, the addition of BAC (biologically activated carbon media) contactors ahead of the filters will not only reduce the organic content, but also will help to reduce the solids loading on the DE filters to permit a more overall efficient treatment arrangement.

Groundwater Treatment

An ideal application of DE filtration would be the treatment of well water supplies. The new federal regulations require filtration of well supplies under the direct influence of adjacent surface water bodies. The particular concern is the removal of cyst and other microbiological pathogens that may migrate to the well. With no pretreatment required for removal, DE filtration can be superior to rapid sand filter processes in the removal of cysts. In rapid sand filter alternatives, the clear well water literally has to be disturbed with the solids of coagulation before it may be filtered.

It is also likely that a DE treatment train may provide a less costly and more simple-to-operate alternative for removal of iron or manganese from well water. The addition of DE and MgO_3 for iron and $KMnO_4$ for manganese, followed by 10 minutes' detention in a contact basin is all that is required before DE filtration. The strong oxidants mentioned before may also be used as optional pretreatment. Without the presence of suspended matter in well water, the DE filter units would be operated at higher rates and/or for longer runs. Two problems may be resolved with DE filtration where iron or manganese may be present in a well supply under the influence of surface water.

Surface Water Treatment

Most of the applications for well water treatment would be for existing supplies not filtered before. Because of long-term regulatory control of surface water, there are few DE application opportunities left for presently untreated supplies. Source water quality permitting, the principal surface water treatment situations for which DE filtration might be considered would include:

— Supplies for new, remote residential or recreational developments serving 200 or more people
— Replacement of outmoded filter plants that require major refurbishment to provide uprated water treatment to meet the current and anticipated standards
— New sources of supply or supplemental supply for the increasing demand of rapidly growing systems.

DE filtration would provide reliable and economical treatment that is simple to operate for the remote, small capacity system. Less-skilled operators may be engaged to run the plant where there is little risk of filter breakthrough and loss of quality. In addition, the complications of disposal of difficult-to-treat wastes from coagulation operations would not exist. Raw water quality permitting, it is probable that another filtration alternative could not provide treatment as reliable at a lower cost and requiring as small a site "footprint" for installation.

Many existing rapid sand filter installations in the U.S. will be forced into major refurbishment programs because of the inability to meet the proposed drinking water quality regulations. It will not be long before the federal requirements stipulate a 0.20 NTU maximum turbidity or even lower for finished water. The basic objective for lowering the maximum limit is to ensure removal of the *Giardia* and *Cryptosporidum* cysts. DE filtration is superior to alternative methods of treatment in removing particulate matter and the cysts included in the 3.0 to 15.0 μm-size range. Where particle counters have been used to determine the effectiveness of solids removal, the DE filter exhibits higher removal than parallel coagulation processes even where turbidity measurements may indicate otherwise.

Rapid sand filter plants that will become outmoded in the future are those with shallow bed filters with limitations in both the depth of media for solids removal and the depth of water over the media that controls the maximum flow rate. In many of these situations, it will be necessary to continue operation of the filter at extremely low flow rates. The hydraulics as well as the structural revisions required usually do not make it practicable to raise the height of rapid sand filters and its building enclosure to increase production. Supplemental filters might be required. A new DE filter installation may prove to be economical for supplemental filters. Where the existing plant cannot be refurbished without major interruption in service, the DE filtration alternative may prove to be ideal. Certain parts of the existing structures may be considered for reuse. Settling basins may be adapted for use as future ozone contact tanks. Existing filters might be converted to future BAC contactors. Chemical feed facilities might be updated to provide space for post-disinfection and corrosion control equipment. Careful review of existing facilities may indicate the possibility of replacing existing filters with DE units at a minimum upset in operation and for an economic cost of improvement.

The third type of DE filter application would be for an additional supply. This supply may be from an existing source where the DE filter plant would parallel the operation of a rapid sand filter plant. The DE plant would be located at a completely new source or at a new intake for an existing source of supply. Where parallel operation will be the mode, it is important not to duplicate facilities in the existing system that might serve the new DE plant. Some operations like waste disposal may be coordinated. When combined, inert DE wastes will enhance the dewatering of the gelatinous hydroxides from rapid sand filter plants. Existing administration, laboratory, and post-treatment could also be modified to handle the new DE filtration operation.

Another attractive feature of DE filtration is its smaller space requirement. This would be helpful to supplement treatment at an existing location where available space for additions is limited.

Design Procedures

The principal steps for preparing a design for new DE filtration facilities are shown chronologically in Figure 7-1. As indicated, there are two parallel routes to be initially implemented, one for unit process and equipment

Background Data	
Review Past Water Quality Data vs. Regs	Inspect and Select Viable Sites Along Transmission Route
Conduct Bench Tests	
Evaluate Efficacy of DE vs. Other Filters	
Review Pressure vs. Vacuum Filter Options	Evaluate Options re Cost and Efficiency of Operation—Zoning and Permit Requirements
Pilot Tests for DE Filter and Supplemental Treatment Criteria	Select Site
Preliminary Engineering	
Plant Arrangement for DE Filter and Ancillary Systems	
Final Design	
Prepare Detailed WTP Plans and Specification Contract Documents	
Construction	
Management	
Preparation of O&M Manual	
Start-Up	
Instruction/O&M Manual	

FIGURE 7-1. Principal Elements of Treatment Project Design

option issues, the other for site location and development requirements. Once the capacity and the type of equipment have been selected and the site and the location of the plant on that site has been resolved, preliminary engineering may proceed.

The site features that involve less technical and more logistic, administrative, and local community issues are discussed briefly here.

These features may at first appear to be secondary to water quality and treatment issues. More water supply programs have been delayed because of community reaction to the apparent impact of plant location on neighborhood values than the type of treatment or even the need for treatment. Unfortunately, the public is more interested in the possible adverse influence of facility location on their individual property values than the health benefits that might be realized with the improved water quality produced by treatment.

Site issues are discussed briefly, followed by reviews of each of the principal steps shown in Figure 7-1. As indicated, this series of steps concentrates first on determining whether DE filtration is a viable process for treatment of a particular source of supply. Once this has been established, the sequence of steps that follow is devoted to developing design criteria and selection of that type of DE equipment and system support that will provide for the most efficient operation. The principal steps to be followed are described here.

The step sequence shown in Figure 7-1 is for a new plant where only the source is known; neither the treatment process nor the plant site has been selected. In many design situations, however, an existing site will be used or water facilities of some kind may have been located before. This may make it possible to eliminate most, if not all, of the parallel site steps so design can proceed once the process and equipment features have been selected.

Site Issues

The three principal considerations in selecting a site for a new water treatment plant are:

— Technical suitability
— Regulatory compliance
— Local acceptance.

The site must have sufficient buildable area to accommodate the plant facilities with enough room for possible future additions. Additional area, either on the plant site or adjacent "open" area, would serve as a buffer zone to reduce the neighborhood impact. The designer is faced with a "chicken and egg" dilemma in this regard. Criteria has not yet been developed to enable sizing the required plant facilities, but site selection should not delay implementation of the overall treatment improvement project. It would be sufficient to use conservative values for sizing the plant "footstep" for site selection purposes. These would include pressure DE filter design at a rate of 2 gpm/sq. ft. of leaf area (1.0 gpm/sq. ft. for the vacuum type) and space for 30 days' storage of bagged DE media. Other accommodations would be provided for DE slurry preparation, waste disposal, and the usual administrative, laboratory, and maintenance areas.

Adjacent or nearby roadway access should be available for convenient delivery of treatment chemicals—notably DE media—and the removal of wastes. Most important, the site elevation should fit the hydraulic grade line from the source to the point of system delivery. This means that the site elevation should not be too high so that excess head would be wasted in the delivery of the treated supply to the distribution system. Conversely, the site elevation should not be so low that head would be wasted in the delivery of the raw water to the plant. Hydraulic conditions will vary. Excess head in the delivery line may be used to operate a pressure filter where directly connected. The same excess head would be wasted in delivery to a vacuum-type DE filter where the top is open to the atmosphere.

Regulatory compliance is related, for the most part, to the local zoning laws and the area environmental restrictions. Even if the site and its proposed development and activity are in complete compliance with the regulations, community objections may generate complicated and extensive environmental impact studies.

The two principal entities that must be satisfied with a proposed plant site are the community groups that benefit directly from the new water supply development and the community group in the immediate vicinity of the site. This latter group may want the treated water benefits, but demand that the plant be located elsewhere to protect property values. Most often, elected officials will support the position of the majority of the group.

Because of the predictable problems that occur with selecting a treatment plant site, more than one technically suitable site should be under consideration where possible. The local community and officials should be included in the initial planning processes so potential objections may be exposed early in the procedures. Once the site has been selected, design may proceed, starting with preliminary arrangement of the proposed facilities on the site.

Meaningful evaluation of alternative sites would not be possible until plant units can be sized to determine site space requirements. This means that site selection activity should first be limited to simple screening of available sites. Referring to Figure 7-1, once the bench tests are completed and the type of DE equipment (vacuum or pressure) has been selected, site comparisons may proceed. Essentially, site comparisons with the attendant technical, suitability, regulatory compliance, and local acceptance evaluations can proceed after the project has been committed to pilot study investigations.

Water Quality

This first phase consists of detailed analyses of source water samples for all water quality parameters included in the federal Safe Drinking Water Act (SDWA) and state regulations. Where possible, analysis data for the source for previous years should be assembled. These data may be available from the water department, from state regulatory sources, from other treatment facilities using the same supply, or from the United States Geological Survey (USGS). Past data will establish maximum and minimum values for critical quality features as well as the length of time and the frequency that maximum levels of contaminants of concern may be expected. These parameters define the operating range of the treatment plant. The water quality evaluations will demonstrate what other quality features besides particulate removal that the DE treatment train may be required to handle.

Bench Tests

The bench tests described in Chapter 5 will help determine whether DE filtration is suitable for particulate removal. These tests will be accurate for only the water sample processed. The background data, particularly for turbidity and color, will indicate how extreme treatment requirements might be for maximum contaminant conditions. Unless maximum turbidity conditions are over the 5.0 NTU level, DE filtration should be able to handle the source water particulates by varying body feed. Bench tests should also determine the need for supplemental treatment, especially for removing THM precursors. Where oxidation using ozone is indicated to improve DE filtration, to provide primary disinfection, or to reduce the THM potential, the use of BAC may be necessary to reduce organics and the distribution system regrowth potential. In essence, the bench tests may not only establish the viability of DE filtration, but also determine those additional process studies that must be included in the pilot tests that follow.

Overall Efficacy

As indicated in Figure 7-1, the advantages of DE filtration over alternative rapid sand filter processes must be evaluated before adoption of the final treatment process. This will involve evaluation of both functional and economic factors. In the comparison with rapid sand filtration, two of the more important features would be:

— Reliability of process and the relative skills and site of the operating staff required
— Disposal of waste—the relatively easy-to-dewater DE wastes vs. the more difficult gelatinous rapid sand filter wastes.

In many instances, the direct filtration mode of the rapid sand alternative will be adequate for the low turbidity wasters suitable for DE filtration. Where the addition of pretreatment for other contaminants to the DE train is necessary, the direct-filtration, rapid-sand option will frequently be the more economical. Direct filtration is, however, vulnerable to breakthrough, even with a skilled operating staff. Direct filtration is also less effective in cyst removal. In these instances, the reliability of DE filtration may be well worth the additional cost for the "peace of mind" it will offer the community. Selection of the process to use should not be based solely on economic features.

Pressure vs. Vacuum Options

Pressure-type filter units are suitable for any diatomaceous earth filtration operation. Even though the vacuum unit must be operated at lower flow application rates on a total capacity basis, it can often be less costly than the pressure filter alternative. The vacuum filter has few mechanical operating features. It is placed flat on the floor. With its open top, all operating phases as well as filter leaf faults may be observed directly. No extra space must be allotted in the installation for the withdrawal of the filter shell or the filter leaf assembly required for pressure type filters. The two principal adverse features of the vacuum filter are:

— Labor-intensive operation in the manual cleaning of filter leaves
— The moist environment and the housekeeping necessary around the units due to evaporation from the open tops and splashing from cleaning.

These disadvantages may be compensated by the fact that less skilled labor is required for operation and maintenance than for the pressure-filter alternative. Generally the vacuum unit would be ideal for smaller-capacity treatment plants and in more remote locations where mechanical maintenance services would be scarce. It is emphasized that, because of the open top, vacuum filters would not be placed in a treatment train directly after ozonation.

Pilot Program

Once the previous steps have been completed, design of the pilot program may proceed. The total program is much more than just planning pilot treatment facilities. In summary, each of the preceding steps contribute the following to the program design:

— Source Quality Data/Regulatory Requirements define the source water contaminants of concern, the variations in the content of these contaminants (max.-min.), the changes over time in contaminant levels caused by seasonal influence or by natural upset like intensive rainfall, and the type of pretreatment and post-treatment that may be required in the treatment train in addition to DE filtration.
— Bench Tests and Filter Options have established the effectiveness of DE filtration and its relative ease of operation (high or low body-feed requirements) and the type filter to be adopted for full-scale design.

Details of the rest of the program design requirements may be found in Chapter 5. These include:

— Design of the DE filter train facilities, including accommodations for pre- and post-treatment (Note: some post-treatments may be accomplished with simple bench tests).
— Data logging that may include intermittent manual entries on data forms, continuously recorded data plotted on charts or entered directly into a computer program, or a combination of these.
— Sample analyses for the significant features of the treatment train (raw water, intermittent treatment, and finished water) to indicate treatment effectiveness as well as providing background data required by the regulations, to be conducted on-site or by outside laboratories, depending on laboratory and personnel staff capacity that may be available.

To be meaningful, the tests and accompanying operating and analysis data must cover all anticipated seasonal and other influences on source water quality as much as it is possible. Some operating conditions like increase in raw water particulates and body feed requirements may be extrapolated to obtain approximate length of filter run. In unusual circumstances, it may be necessary to leave the pilot facilities in place for a time in preparation for operation in periods of high runoff that produces high turbidity and/or color or in periods of high algae proliferation. The magnitude of the impact of these unusual circumstances on DE filtration will determine the need for the additional test data.

The final product of the pilot program should be a report in sufficient detail to satisfy the regulatory authorities and to provide the background data needed to facilitate design.

Preliminary Engineering

The shape of the new treatment design is developed in the preliminary engineering step. Alternatives are evaluated, not only for principal items of equipment, but also for the ancillary systems that are part of the DE filter treatment train. In essence, the most important functional design decisions are made in this phase of design. Once completed, the preliminary designs form the basis for the preparation of the contract plans and specifications that are used for the soliciting of competitive bids and for the construction of the proposed treatment improvements.

The type of DE filter will have been decided before the pilot tests so the filter options would be those related to pertinent operating systems and controls. Where relevant, design options would be evaluated for pretreatment before the filter step; for DE media delivery, handling, and preparation; and for the disposal of the sluiced wastes. These options may vary considerably by location and by plant operating needs. For instance, maintenance services for a plant located in remote rural areas might not be readily available. The design would, therefore, concentrate on the use of system requiring more manual operation with a minimum of automation and complicated mechanical equipment. The same capacity plant may be located near industrial areas where instrumentation and mechanical maintenance personnel are common in the labor force. For these situations, such maintenance capability may even be added to the plant staff to permit adopting designs that provide for more efficient operation with reduced personnel requirements.

Preliminary engineering may be divided into three individual phases as follows:

1. Development of background design data
2. Evaluation of design options
3. Arrangement of the selected equipment and system options on the plant site.

The three procedures are discussed here. The salient features of an assumed treatment situation are used to illustrate a typical design.

Background Design Data

The elements of base data defining the treatment plant design needs are shown in Table 7.1. System water supply demand parameters and the essential pilot tests data are used to calculate the basic factors used to develop major equipment, system, and building enclosure requirements. These data provide not only the information for determining equipment capacity and a basis for selecting the type of equipment or system to use in design, but also the spatial needs for building size and arrangement.

For clarity, a design situation has been assumed, the principal test data and supply needs of which are shown graphically in Figure 7-2.

TABLE 7.1. Base Design Data

1. From Pilot Tests
   - Body feed requirement variations vs. turbidity
   - Length of run vs. filter application rate and body feed
   - DE media-grade precoat and body feed
2. From System Needs
   - Average annual supply demand (existing-projected)
   - Maximum and minimum month supply demands
   - Maximum day demand.
3. Calculated
   - DE media requirement variations, storage needs
   - DE filter unit application rate and production demands, number of filters
   - Plant water needs for operation
   - Waste flow volume and solids.

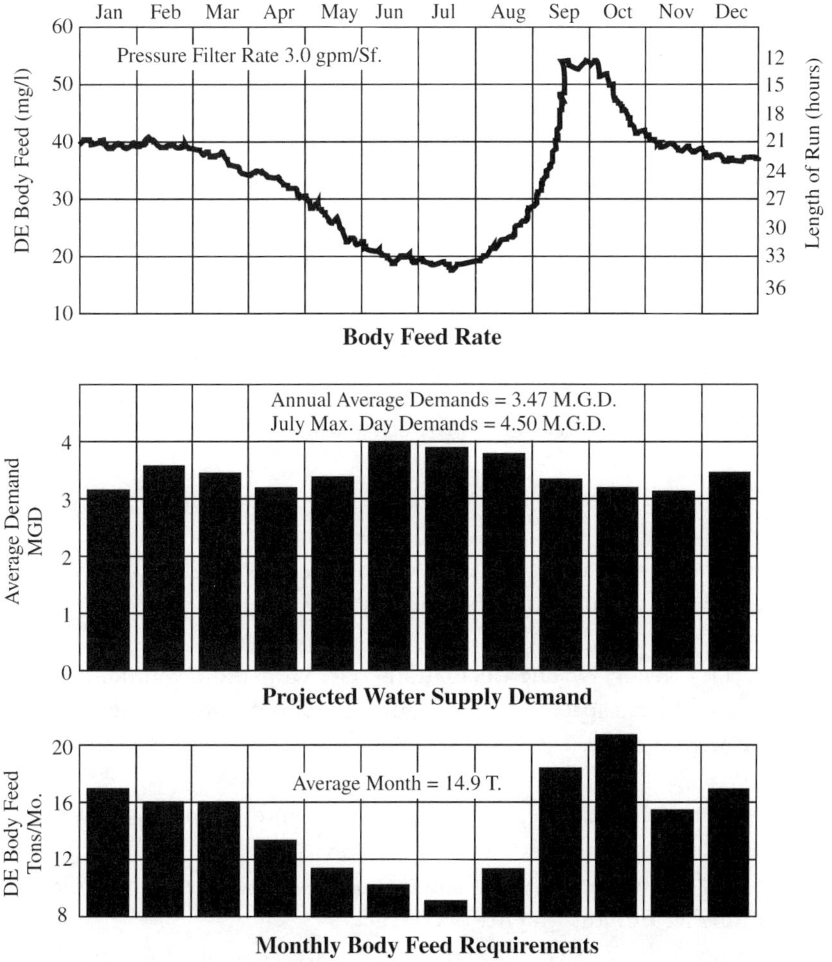

FIGURE 7-2. Monthly Variations

As indicated, the pilot test data has provided the body feed requirements for 12 months' operation and the attendant length of filter runs for operation at a leaf application rate of 3.0 gpm/sq. ft. These are data from an actual pilot test run in which the length of run happened to be indirectly proportional to the body feed rate. This meant that raw water solids characteristics were reasonably consistent through all seasons of operation. It is cautioned that in other locations with influences like algae growth or changing stream flows, raw water solids features may vary significantly, having more influence on length of run patterns. For these situations, length of run would simply be handled separately, independent of body feed needs.

For simplification of presentation, the data is shown as monthly averages. For certain situations, it might be more advantageous to address maximum and minimum conditions by the particular period of time rather than the calendar month. Average annual water supply demand for the example in Figure 7-2 is 3.47 mgd. This should be the demand for a projected period of time 10 to 20 years in the future. Fortunately, for this design, maximum day and monthly average demand does not occur simultaneously with the fall reservoir turnover when length of run is at a minimum. Using the data in Figure 7-2, body feed demands may be translated to monthly delivery requirements using water supply demand and average body feed rate in the calculation:

$$\text{DE (lbs)} = \text{mgd} \times \text{mg/l} \times 8.34 \times 30 \text{ (days in month)}$$

The example calculations, based on the background data outlined in Figure 7-3, are discussed later by category.

FILTER CAPACITY. The calculations start with the determination of the required filter capacity or filter leaf area. As indicated, the initial calculation steps include:

— Determine leaf area at the test rate for minimum length of run conditions
— Calculate approximate length of run for projected conditions of higher turbidity, higher body feed
— Since the run is less than minimum acceptable cycle of 12 hours, adjust filter application rate to ensure more than 12 hours of operation for all anticipated conditions
— Determine filter leaf area for adjusted rate
— Adjust leaf area for 45 minutes out of operation for sluicing (this is the area used to specify filters).

As indicated, 990 sq. ft. of filter-leaf area would be required for continuous operations under maximum body feed conditions.

DE MEDIA NEEDS. Calculations continue with determination of the maximum month supply of DE media for body feed and precoat as follows:

— Body feed: monthly weight from Figure 7-3, reduced to daily need
— Precoat: daily weight based on cycles/day and unit application rate
— Daily requirements: (total of the above) reduced to equivalent bags of 50-lb and 900-lb capacity
— Net area required to store each of the alternative capacity bags for 30 days' needs
— Bag handling under minimum DE media requirement conditions.

The aforementioned data is used in the planning of building facility options and plant operation logistics as well as in the selection of the type of bag storage to adopt. Thirty days' storage of treatment materials is a common requirement in many state standards. The bag handling requirements, especially for the relatively short period of maximum need, can be helpful in planning a plant staffing approach. It might be possible, for instance, to supplement the plant labor staff somewhat for the increased handling of 50-lb bags during the period of maximum DE use shown in Figure 7-2. Alternative use of 900-lb bags in this design situation would require less labor for bag handling under all conditions of media need, but with more costly bag handling equipment accommodations. Under maximum DE use conditions, 35 5-lb bags or two 900-lb semi-bulk bags/day would be used under maximum conditions. Usage is reduced to 15.5 50-lb bags/day or less than one 900-lb bag a day under minimum conditions.

Media Storage Area. In the next step in Figure 7-3, the net 30 days' storage area required for the two types of bags stacked on pallets is determined for maximum use conditions. This area provides the base for planning overall DE storage, which also includes space for unloading equipment and aisles for forklift transfer of pallets.

Sluice Wastes. The total volume of sluiced wastes discharged daily is the total of the DE media used in filtration and the amount of water used in the spray operation. As shown in Figure 7-3, for the example, this would be the DE daily use calculated before added to the unit rate of spray multiplied by the leaf area and the length of time for cleaning. Water volumes have been based on a unit spray application of 0.3 gpm /sq. ft. of leaf area applied for 10 minutes. As a word of caution, each type of DE filter design will have different rates and wash procedures. Perhaps the most efficient operation would be the manual washing of vacuum type filter leaves using a single high-pressure spray. Here, little water is

Test Data (From Fig. 7-1 Procedures)

- — Filter type: pressure
- — Filter application rate (test): 3.0 gpm/sq. ft.
- — Minimum run (@ 54 mg/l body feed): 13 hours
- — Projected demand @ minimum run (Figure 7-2): 3.33 mgd (2330 gpm)

Calculations

FILTERS.

— Projected minimum run @ 65 mg/l body feed (assumed max.): (54 ÷ 65) mg/l × 3.33 hrs. = 10.8 (<12)

— Projected filter application rate (12 hrs @ 65 mg/l): 3.0 gpm sq. ft. × (10.8 ÷ 12) = 2.7 gpm/sq. ft. For design, use 2.5 gpm/sq. ft. max. rate

— Total Filter Leaf Area
@ Minimum run (Sept.–Oct.), demand: 2330 gpm
2330 gpm × (1 × 2.5 gpm/sq. ft.) = 932 sq. ft. total
Adjust for sluice/precoat time: 45 minutes
(12.75 ÷ 12) hrs. × 9.32 = 990 sq. ft.
Maximum day check (Figure 7-2)
4.5 × 700 × (1 ÷ 3.0 gpm/sq. ft.) = 1050 sq. ft.
Adjust for sluice/precoat time
(33.75 ÷ 33) hrs × 1050 = 1073 sq. ft.
Maximum day controls: use 1050 sq. ft. (rounded to 1000 sq. ft.)

DE MEDIA USAGE.

Maximum Month

— Body Feed, max. month (Figure 7-2) = 20.5 tons
20.5 tons × (2000 ÷ 30) = 1366 lbs/day

— Precoat, max. cycle = 12 hrs.
990 sq. ft. leaf area × 0.2 lbs/sq. ft. = 198 lbs/cycle
198 lbs/cycle × twice daily = 396 lbs./day (approx. 400 lbs)

— Total DE Daily Use (max. conditions)
1366 + 400 = 1766 lbs/day

— Daily DE Bag Handling: Options
50-lb bags = (1766 ÷ 50) = 35 bags
900-lb bags = (1766 ÷ 900) = 2 bags

FIGURE 7-3. Typical Design Calculations *(continued on next page)*

Minimum Month (9.1 tons)

50-lb bags = (9.1 ÷ 20.5) × 35 = 15.5 bags/day

900-lb bags = (9.1 ÷ 20.5) × 2 = 0.9 bags/day

DE STORAGE (30 DAYS).

— 50-lb bags, 3 bags/layer × 4 layers = 27 per pallet, 3′ × 5′ × 6′ high
Storage area = [35 bags × 30 days × (3′ × 5′)] ÷ 27 = 583 sq. ft.

— 900-lb bags, 1 per pallet, 4′ × 4′ × 4′, stacked two pallets high
Storage area = [2 bags × 30 days × (4′ × 4′)] ÷ 2 = 480 sq. ft.

SLUICE WASTES.

— Maximum month water volume
Solids: raw water solids, 2.5 NTU × 0.75 = 1.9 mg/l
1.9 mg/l × 3.3 mgd × 8.34 = 52 lbs/day
DE usage (lbs/day) = 1,766
Total = 1,818 lbs/day
Sluice water (0.3 gpm/sq. ft. of leaf area)
0.3 gpm × 1000 sq. ft. × 10 minutes = 3,000 gallons
Sluice twice daily = 6,000 gallons
Sluice water weight × 8.34 = 50,040 lbs
Total weight = 1818 + 50,040 lbs
% Solids × [1818 ÷ (1818 + 50,040)] × 100 = 3.5
Total volume = (1818 + 50,040) ÷ 8.34 = 6218 gallons/day (8.30 cu. ft.)

— Minimum month water volume
Filter cycle = 34 hours
24-hour day ÷ 34 = 70% of filter area/day
[6000 gpd (max.) ÷ 2 (daily)] × 0.7 = 210 gallons/day
Roughly 1/3 of maximum

FIGURE 7-3. *(continued)*

wasted, as the operator can observe the entire procedure. Pressure filters with rotating leaves would use less water than fixed leaf designs with oscillating sprays. Both the rate of spray application and the average time for cleaning that should be expected would be provided by the manufacturer of the respective filter units. As indicated, the waste volume for maximum washing conditions (shortest runs) has been calculated as 6,218 gallons a day. Average concentration of DE plus raw water solids removed by the filters would be around 3.5%. For minimum month (longest run) conditions, the total waste volume is reduced to 4,200 gallons, or about 70% of the maximum. It is noted that this latter calculation assumes that all filters would remain in service as the length of runs are extended. Another operating option exists where part of the filters could be inactivated, which would shorten runs and increase washing cycles for the reduced units remaining in operation.

Diatomaceous Earth Filter Options

For the design example, the decision to use pressure filters was made in the pilot stage of the design program. The calculations (Figure 7-3) indicate selection of a 3.0 gpm/sq. ft. filter application rate with a total filter leaf area of 1,100 sq. ft. available for maximum system supply demand conditions. Spare capacity must be added in case one filter is out of service for mechanical maintenance. The greater the number of units used, the smaller the overall capacity required. For instance, 1,100 sq. ft. of filter area may be divided up as in Table 7-2.

Prices must be obtained from the manufacturers to determine the least cost. Depending on the type of DE unit selected, the four units (one a spare) rated at 350 sq. ft. each could be less costly than the other options despite the greater overall leaf area.

The other multiple unit features that should be considered are as follows:

— It would take less floor space for installing four 350 sq. ft. filter units than the five and six filter options.
— For the capacities under consideration, operation and maintenance of individual units would be relatively the same; i.e., six filters would involve about 50% more personnel time than four larger capacity units.

For the example design, the selection would be four 350 sq. ft. filter units depending on the type of DE filter. The different types of DE filter

TABLE 7.2. Sample of DE Filter Options for 1,000 Square Feet of Area

No. of Operating		Leaf Area	
Units	Operating Unit	Added Spare	Total
3	350	350	1400
4	263	263	1315
5	210	210	1260

units available were discussed in Chapter 4. Three types of pressure filters would be suitable for the individual 350 sq. ft. capacity as indicated in Figure 7-4. The significant operating features of each type unit are noted in the Figure. While more costly, the rotating leaf filter would require the least installation space and could be operated more efficiently. Maintenance of filter leaves would also be more costly because individual filter leaves must literally be pulled from the center shaft for maintenance.

The vertical unit offers the options of manual sluicing from the top or automatic sprays, but requires an elevated platform for access to the unit. Moreover, the odd-sized filter leaves are somewhat difficult to pull and reinstall because the connection socket is at the bottom of the tank shell.

The horizontal retracting-shell DE filter requires the most floor space. When the shell is retracted, however, the filter leaves are easier to inspect and to remove for maintenance. From a labor skill and number standpoint, the three units may be compared as follows:

— *Vertical Unit.* Fewest moving parts. Will have the least maintenance and require the least skills for both operation and maintenance. If the units are manually cleaned, additional labor may be required.
— *Horizontal Unit.* Requires routine maintenance for rotating spray seals (through shell) and for motor-operated retraction device. Mechanical maintenance skills should be available on staff or from nearby contracted services. This unit may also be computer-controlled for which on-site skills would be needed.
— *Rotating Leaf.* Requires on-site or contracted mechanical skills for rotating shaft device and for pulling leaves for inspection and maintenance. May use computer controls.

If the vertical unit has installed sprays, the size of the on-site operating staff would be about the same for all three units. The vertical units would require relatively unskilled labor for operation and maintenance.

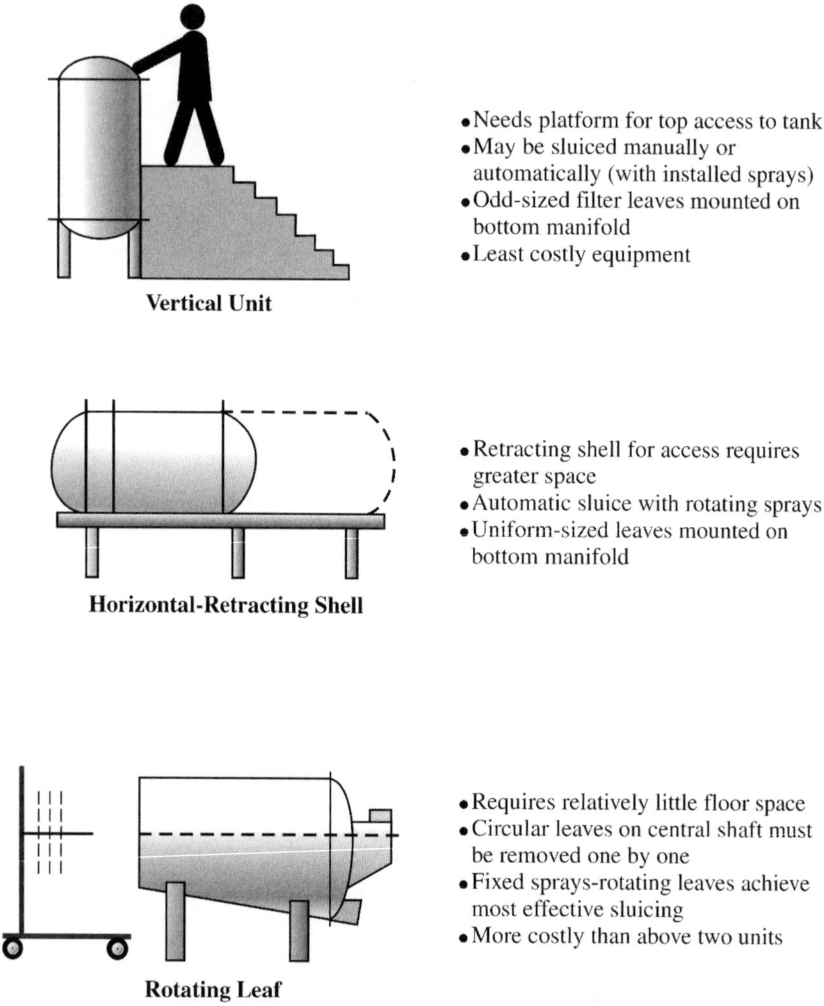

FIGURE 7-4. Diatomaceous Earth Filter Options

The horizontal unit would require a somewhat more skilled operating staff, especially if the unit operation is computer controlled. Mechanical maintenance capability should be available on staff or from convenient nearby contractor mechanical services.

In summary, staff comparisons are related more to required skills and attendant salaries than numbers. The rotating leaf filter should have the highest salary base, the vertical option the least.

Diatomaceous Earth Delivery, Handling, and Storage Options

As indicated in the Figure 7-3 calculations, use of DE in the example design situation would range from 15.5 to 35 50-lb bags or 0.9 to 2.0 900-lb bags of DE media a day. Referring to Figure 7-2, it appears that the maximum or close to maximum use would last within a two-month period of time each year. It is also noted that source water quality conditions requiring the maximum rate of DE use indicated for the sample situation do not necessarily occur every year. In this situation, it probably would be more meaningful to compare the two options on the average month basis, which is 14.7 tons or about 970 lbs/day. This means that daily handling would involve about 19 50-lb bags and just 1-900 semi-bulk container.

As indicated in Figure 7-5, the selection of which type bag to adopt for design would depend upon the cost of equipment and labor of operation. Space accommodation should also be included. As indicated in the calculation (Figure 7-3), 30 days' storage for maximum use would be about 480 sq. ft. for 900-lb bags and 100 sq. ft. more, or around 580 sq. ft. for the smaller 50-lb bag. Equipment for the 900-lb bags would be more costly. A heavy-duty, high-lift forklift would be required for both unloading delivered bags and transferring bags to the discharge rig. This rig is also an addition in cost for the larger bags.

In any case, the 900-lb bags must be emptied into a single, concentrated slurry tank in order to maintain control of the percent concentration. Otherwise, a weigh feeder would be required to divide the bag contents between tanks. At 900 lbs and a 15% slurry concentration, about a 700-gallon tank would be required, or probably a 1,000-gallon tank to allow for emptying the 900-lb bag into a partially filled tank. The concentrated slurry would then be distributed to day tanks for dilution and use. On the other hand, multiple smaller-capacity tanks may be used to receive the daily 50-lb bag requirement.

For the sample design, the daily DE quota would be divided between three active day tanks. Assuming 3% slurry concentration, the day tank capacity would be:

$$\frac{970 \text{ lbs/day}}{3 \text{ units}} \times \frac{1}{0.03} \times \frac{1}{8.34} = 1{,}300 \text{ gallons}$$

Using a 5-foot diameter day tank, the fill height would be about 9 feet, total height about 10 feet. It is noted that the three tanks hold the amount

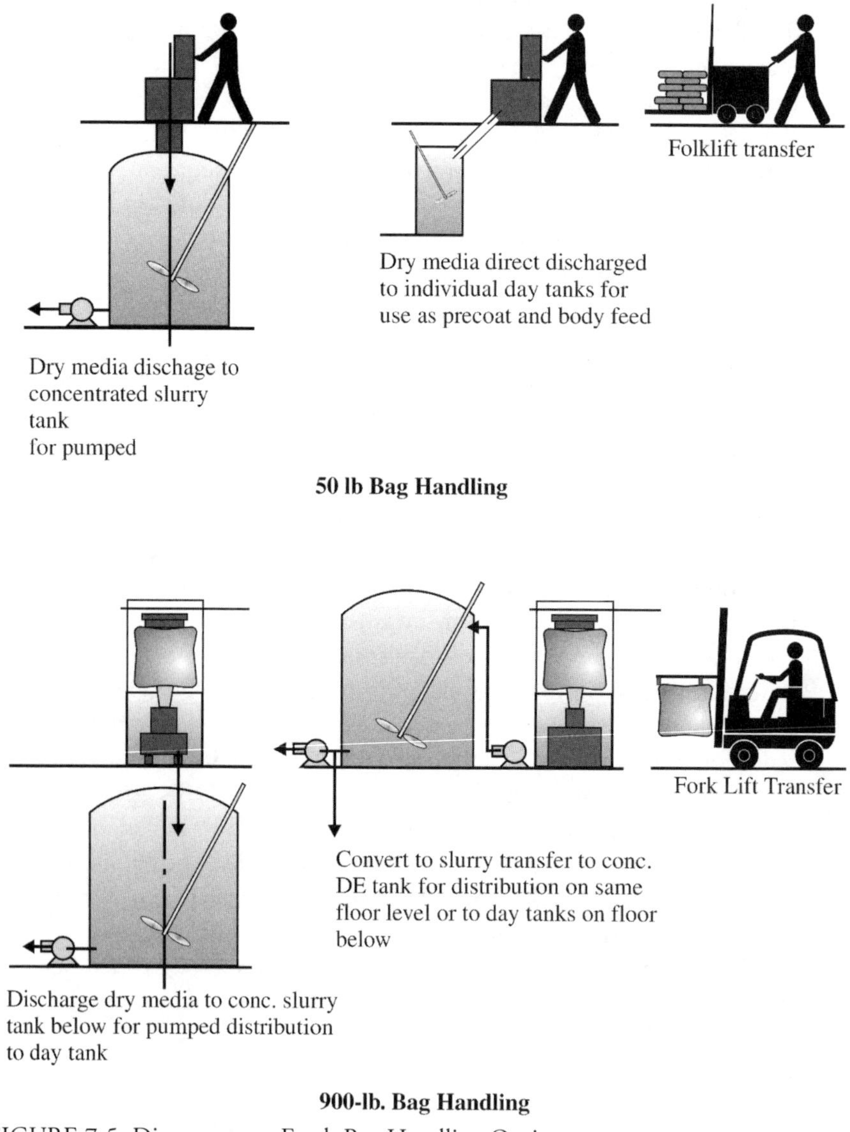

FIGURE 7-5. Diatomaceous Earth Bag Handling Options

of DE required for precoat as well as body feed under average conditions. If a separate precoat tank were to be furnished for the three operating filters, the body feed tank would be less than 1,300 gallons.

The day tanks would be filled with concentrated slurry and diluted with the respective filter out of operation after a sluicing cycle. As shown in Figure 7-5, 50-lb bags may also be emptied directly into the day tanks for slurrying. Depending on the skill of the operators, 19 50-lb bags could be emptied along with the usual housekeeping (disposal of bags, sweeping spills) in from 1.5 to 3.0 hours by a single operator with a motorized forklift. It might take two operators from 30 minutes to an hour to empty a single 900-lb bag with the usual cleanup after. In other words, more labor for handling 50-lb bags, but with less sophisticated equipment and allowing for a direct discharge to a day tank. The selection would be based on the comparison of a more labor intensive system vs. a system that uses more costly, sophisticated equipment.

Waste Slurry Disposal

Details of waste handling options were discussed in Chapter 5. The principal parameters or features that would be considered for alternative approaches to DE waste disposal are outlined in Table 7.3. In essence, those features that would be common to any method of waste disposal include the following:

— Free gravity discharge of sluiced wastes from individual filters to a holding basin on a level below

TABLE 7.3. Diatomaceous Earth Waste Disposal Parameters

— Receiving basin for sluiced wastes should be located on floor below to permit gravity discharge from each filter.

— Waste slurry must be concentrated by decanting or by conditioning and settling separation before dewatering.

— Either vacuum filters or filter presses may be used to dewater thickened, conditioned DE waste slurries.

— Supernatant from decanting, settling/concentration or from dewatering may be recycled if permitted by the pertinent state regulations and with treatment in accordance with those regulations.

— Dewatering of all waste solids or those remaining after DE recovery for disposal as a solid waste
— Disposal of both waste solids and waste supernatant off-site as well as use of recovered DE for body feed and recycling of supernatant in strict accordance with the pertinent state regulations.

All DE filters are fitted with large-site bottom drains to allow free discharge of the sluiced flow. The hydraulic conveyance of that discharge to a holding basin should be designed so as not to inhibit that flow.

The state (or federal where no primary exists) regulations will define both reuse and disposal of sluiced DE wastes. Depending on the particular rules, both reused DE and recycled supernatant may have to be disinfected to avoid recycling of live pathogens that had been previously separated in filtration. Supernatant discharged off-site may be adequate in quality for connection to the local sewerage system. Supernatant discharged to adjacent surface waters or for leaching to groundwater aquifers must be provided treatment in accordance with the pertinent regulations.

While detailed design of waste facilities are beyond the scope of this text, a few rules of thumb are pertinent to any design:

— Base holding capacity of receiving basins must be equal to the discharge expected under operating conditions of highest body feed use and most frequent filter cleaning cycles.
— Additional holding basin capacity must be provided to allow for:
  - Decanting supernatant after the solids have been allowed to settle, usually for a settling rate of 0.000082 ft/sec or about 7 ft/24 hrs for 90% of the normal Hyflo particle fraction
  - The periods between dewatering or recovery operations which are most often on a day shift and less than daily basis
  - Possible breakdown and maintenance of waste processing facilities with limited redundancy.

All holding basins should be provided with mixers to maintain the slurry in suspension. Heavy-duty mixers should be provided to permit starting with a settled slurry that has resulted from decanting or accidental mixer shutdown.

For the example design, maximum day sluiced waste discharge would be about 620 gpd (Figure 7-3). Assuming no waste processing over the weekend, three days' holding capacity would have to be provided or

18,600 gallons. With around a 10% contingency factor, the holding basin volume is rounded off at 20,000 gallons. If decanting is to be practiced, it would be best to provide three 700-gallon basins, each of which would receive a single-day sluiced waste. The basins might be sized as follows:

— Water depth = 10 ft
— 700 gallons × (1/7.48) = 93.6 cu. ft
  700 gallons × (1/10 ft) = 93.6
— Say three 10 ft × 10 ft × 10 ft depth basins
— If decanted 5 ft = (10/5) × 3.5% = 7% concentration

With decanting, unless discharged to a nearby sewer or to an outside holding lagoon, three days' supernatant must be held for further treatment for reuse or other purposes. A more useful design would provide the following:

— Three 700-gallon basins for receiving and decanting
— One 1,100-gallon basin to receive supernatant

A 1.0 ft × 10 ft × 11 ft water depth basin would hold supernatant with no mixer necessary. Only the very lightest of the waste particles would be carried over in the supernatant. The hydraulic action of fill and withdrawal would stir up any particles that might settle for removal. Where possible, the basins in the lower level of any treatment facility could be constructed of concrete and become part of the building foundation. Discussion of waste facilities is not discussed beyond this point.

Site Arrangement

The basic rule for the arrangement or layout of water treatment plant facilities on a site is to take full advantage of the topography for the flow of operation and the natural tree cover to shield the plant from its environs. Treatment plants are rarely located in areas of industrial development. Most often they are located adjacent to water bodies of good quality or somewhere en route from the source to the point of use of treated water. It is especially important to shield the industrial features of a plant from residential view. Such features would include truck delivery of treatment chemicals and materials and the parking of staff vehicles. Beyond this, it is important to use site relief to provide for more efficient movement between operations of a plant as well as for economical building design.

Addressing all or even a major portion of the ramifications of adapting a DE treatment plant arrangement on the many variations of site topography that may be encountered would be difficult. The discussion here, therefore, has been limited to the preferable relationships between the principal DE plant systems in locations on sites of flat or sloping characteristics. The principal relationships to consider include:

— DE bags should be unloaded at the rear of a treatment plant building or at another location shielded from the main access roadway.
— Personnel parking should be in a shielded location.
— The dry DE bag storage and handling area should be separated from the rest of plant by enclosure with outside access to avoid "tracking" dust.
— If possible, dry DE emptied from bags should flow directly into slurry mix tanks partially filled with water.
— Sluiced wastes should be discharged by gravity to holding basins located on the floor below.

To summarize, separated dry DE areas should be located above slurry mix tanks and DE filters above waste holding basins if possible. This suggests a three-level arrangement with the DE filters and slurry tanks on the main floor, dry DE storage and handling on the floor above, and waste handling on the floor below grade. There may be situations where such ideal arrangement is not feasible or economical. Here, the dry DE area is separated by partition on the same floor as the DE filters. The bag discharge is immediately mixed with water and pumped to adjacent holding tanks. This arrangement was discussed in the previous chapter for 900-lb bag handling. For 50-lb bags, bag breakers and dust collectors would have to be located on a platform slightly elevated above the water mix tank on the floor. Pallets would be deposited by forklift on elevated platforms for easy access or the pallet hoisted by forklift can simply be moved to the necessary location for easy lifting of individual bags by the operator.

Three cross-section arrangements are shown in Figure 7-6 for flat sites. The bi-level (A) arrangement would have DE delivery and handling on the same floor as the DE filters. This floor is elevated sufficiently above grade to permit tailgate unloading from delivery trucks at the rear of the building.

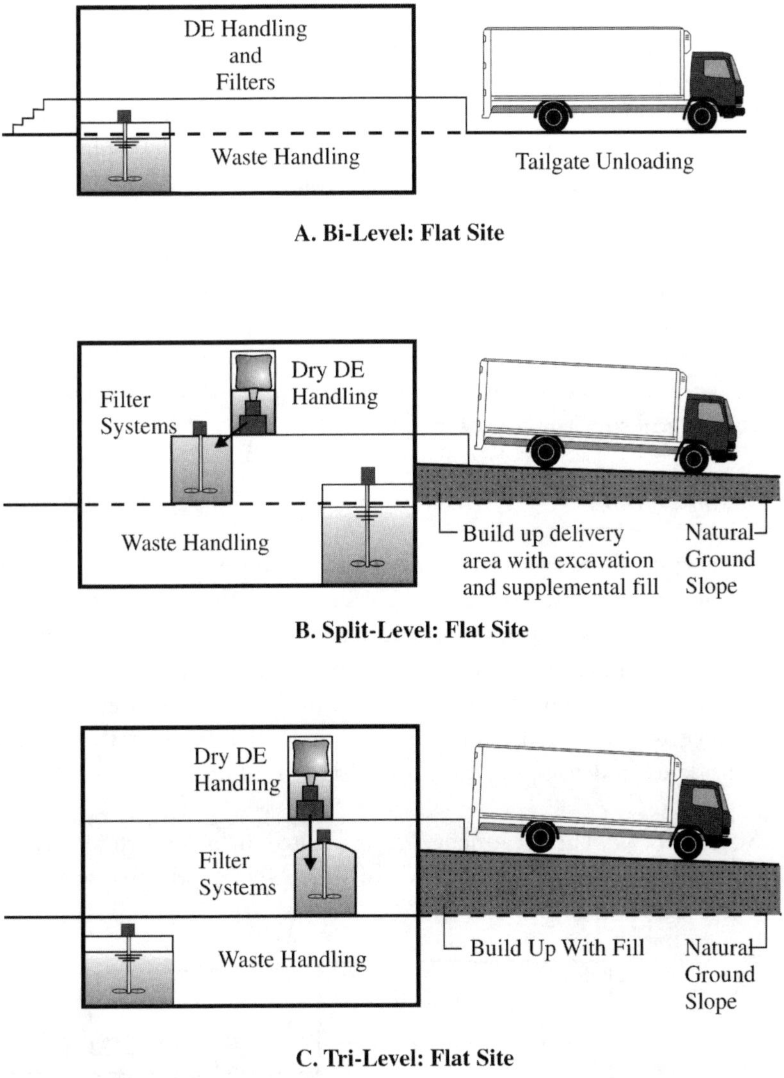

FIGURE 7-6. Flat Site Arrangement Options

A considerable amount of earth would have to be excavated to pour the concrete foundation of the building to the level required for waste handling basins. As indicated, this excavated material may be used to build up the grade behind the building to create the split-level (B) and tri-level (C) arrangements on the Figure. The excavation may provide sufficient material for the split-level fill. Supplemental fill may be required to build up the rear grade for the tri-level site. Grade adjustment for these two arrangements includes that required for the roadway that would be gradually elevated from front to back around the building.

Figure 7-7 illustrates how cut and fill in a sloped site may be used to adapt the three-level, DE plant arrangement. It is noted that locations on very steep sites may require unloading at the side of the building to limit excavation.

Numerous arrangements are possible. It is only important that the designer take full advantage of the site features to develop efficient and economical site arrangements.

Selection of Options

Reviewing all the situations that might influence selection of the options for final design discussed in the previous example is impossible. Two site location scenarios have, therefore, been chosen to illustrate the influence local situations should have on design of treatment plant facilities. Two deliberately different situations provide the basis of selection of options to emphasize some of the more important local influences that should be considered in developing meaningful design solutions.

Scenario X. The treatment site is located in a semi-rural area where homes are separated by hundreds of feet. Average property size is greatly

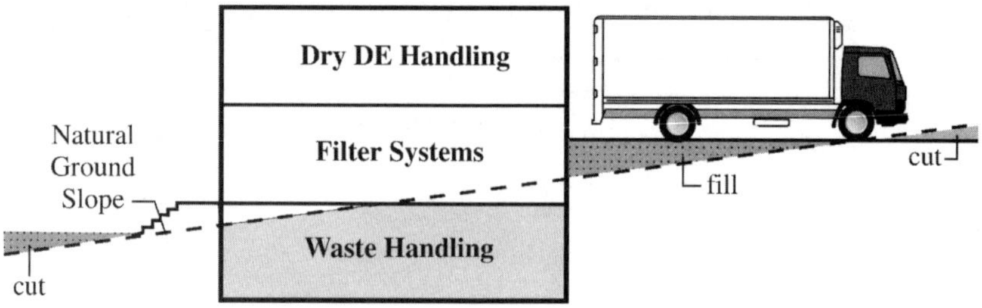

FIGURE 7-7. Sloped Site Arrangement

in excess of that required for plant arrangement. Local labor available to staff the plant is plentiful, but relatively unskilled.

The two significant options discussed before were the type of DE filter design and the bagged DE media for use in operation. Plant design selections for these alternatives would be:

— Vertical Filter, with the least mechanical features to maintain and with the possibility of manual washing to ensure clean filter leaves with a reduced cost of filter unit
— 50-lb DE Bags, the unloading and handling of which would be more labor-intensive and requiring more space for storage than the 900-lb semi-bulk bag alternative.

The pool of unskilled labor available at lower base salary rates would compensate for the smaller, more skilled plant staff required for operation of the horizontal-retracting shell and rotating leaf filter alternatives. The larger staff would also be able to handle the more labor-intensive handling of the 50-lb bags. The use of 50-lb bags and the larger staff would also permit charging individual precoat and body feed day tanks. The distribution of concentrated DE slurry necessary in the use of 900-lb bags is a more costly system with more mechanical, piping, and pumping features. The large separation of properties and the large available site area would minimize the possible problem of the larger building area required to store and handle the 50-lb bags.

SCENARIO Y. A treatment site of limited size is situated in an essentially residential area where community activists against the location of the plant will settle only for a plant with a layout of "small footprint." The plant would be close to the community served which has sufficient industrial development supporting numerous local mechanical and electrical contract services.

Here, there is practically no choice of options. To produce the smallest possible "footprint," the selection would include:

— Rotating leaf filter
— 900-lb semi-bulk DE media bags.

The higher skilled maintenance required for the filter and DE handling systems could be provided either by staff here or outside contract services.

SUMMARY. The selection of options must be influenced by the local conditions, which may include:

— Community acceptance
— Site buffer potential
— Capital funding available
— Available labor pool (number and skills)
— Available contract services
— Staff salaries and maintenance permitted by operating budget.

Public acceptance is paramount in regard to both plant location and project costs. The economic ground rules should be established early in a project to provide a guide for selection of options and to avoid unnecessary design efforts that may not be approved. Site selection should also be guided by the potential acceptability if alternative locations may be considered.

Contract Documents: Construction

The design procedure presentation was completed in the Preliminary Engineering Section. The evaluation and selection of most of the major process equipment and system options is accomplished in this previous step. The Contract Document step consists of preparation of the final design plans and specifications in sufficient detail to provide a meaningful basis for competitive bidding by construction contractors as well as for state and local review to obtain the required approvals.

The design details and the appurtenant specifications for major process equipment and systems should allow sufficient flexibility to allow use of all viable manufacturers' equipment. This means that space allowances and piping details in the Contract Documents should not be restricted to single manufacturer's offerings.

After the contract is awarded to the successful bidder, the engineer will review and approve shop drawings for the equipment offered. Once the approvals are completed, operation and maintenance manuals for all principal equipment should be collected to be used in the preparation of the Plant Operation and Maintenance Instructions. This manual provides the background for training the plant staff as well as guidance for start-up and troubleshooting operations. The plant design is complete with the issuing of these instructions along with the manufacturers' files on operation and maintenance.

START-UP/TROUBLESHOOTING OPERATION

The basic objective in the operation of a DE filter is water of low clarity produced in the longest filter run practicable. As discussed in the Pilot Testing chapter, there are limited functions in the operation of a DE filter that may be controlled. In summary, the filter operation should be responsible for the following:

— *Precoat.* Charge sufficient virgin DE to the recirculation flow to deposit the proper precoat depth, ensuring that the recycle stream runs clear before completion of the cycle.
— *Filtering.* Switch the valve controls manually or automatically from the recirculation to the filtering flow-through mode; set the rate of body feed to attain the most efficient length of run.
— *Sluicing.* Stop filter operation when maximum loss of head is reached and initiate draining and sluicing the cake from the filter leaves; thoroughly clean filter leaves manually or through proper setting of timing controls.

The principal errors committed manually or in the setting of automatic controls include:

— Improper rate setting for filter flow, charging precoat, and adding body feed
— Improper timing in the switch from the recirculation (precoat) to the filtering mode of flow
— Incomplete sluicing of filter leaves.

For the most part, DE filter operation is a fail-safe system in regard to particle breakthrough. Even fail-safe systems, however, are not absolute in operation. Table 7.4 is a list of the specific operating errors and mechanical system faults that may cause loss of product water clarity or breakthrough. As indicated, achieving clarity is an objective in both the precoat

TABLE 7.4. Troubleshooting Insufficient Clarity

	Operating Cycle	
	Precoat	Filter
Operating Error		
Insufficient DE feed	X	
Recirculation rate too slow	X	
Filter flow rate too fast (wash cake)	X	X
Dirty screens	X	X
Precoat slurry too dilute	X	
DE grade too coarse	X	X
Improper switching of valves		X
Flashing of filtrate at septum (cold water/high flow rate)		X
Filter air bound	X	X
Air sucked in through precoat tank vortexing	X	
Mechanical Faults		
Filter not properly vented (partially filled with air)	X	X
Poor baffling - cake washed off	X	X
Septum weave too open	X	X
Seal leaks - screen connection	X	X
Warped screens		X
Septum too loose (flexible)		X
Dirty screens	X	X

and filter cycles for which the problem features are indicated. In addition to the operator responsibilities listed previously, loss of clarity may be caused in vacuum filter operation with cold water when too high a flow rate may produce negative pressure in the cake to release entrained air. Improper filter design may produce vortexing of inlet flow due to poor baffling or filter leaf exposed to air with poor venting. Other faults are due to improper leaf design and too coarse a grade of DE for operation. Most of the problems affecting product water clarity (filter) in Table 7.4 would occur first for precoating. The filter cycle should not be started unless the recirculation flow runs clear. This means that the list of potential problem factors affecting breakthrough alone is reduced to a small number. By eliminating mechanical faults, breakthrough alone would be caused by improper switching of valves and cold weather operation.

Table 7.5 lists the most common errors and faults that would produce shortened filter runs. Most of the faults listed would not be common to a well-designed filter and filter system. These errors would easily be avoided by observing the simple rules of good operating practice.

TABLE 7.5. Troubleshooting Short Filter Runs

Operating Errors
— Too high application rate
— Too high or too low body feed rate
— Erratic body feed rate—plugged pump
— Improper combinatoin of precoat and body feed DE grades
— Precoating with unfiltered water
— Excessive recirculation during the precoat cycle.

Mechanical Faults
— Insufficient feed pump capacity
— Back pressure too high
— Flow lines too small
— Poor leaf drainage
— Pump sucking air (through seals)
— Erratic body feed pump.

References

American Water Works Association (AWWA). 1993. *Precoat Filtration.* Manual M30, 2nd ed. Denver, Colorado: AWWA.

Bauman, E.R., Cleasby, J.L., and LaFrenz, R.L. 1962. "A Theory of Diatomite Filtration." *Journal of the American Water Works Association,* 54 (September): 1109.

Black, H.H., and Spaulding, C.H. 1944. "Diatomite Filtration Developed for Field Troops." *Journal of the American Water Works Association,* 36 (November): 1208.

Johns-Manville. 1975. *Dry Cleaners Handbook,* 11th ed. Lompoc, California: Johns-Manville.

McIndoe, Jr., R.W. 1969. "Diatomaceous Earth Filtration for Water Supplies." *Water and Wastes Engineering,* October: 50–52; November: 49–52

Index